T0282566

EROSION and SEDIMENT CONTROL for RESERVOIR SEDIMENTATION from AGRICULTURAL ACTIVITIES in HIGHLANDS

PROF. LARIYAH BTE MOHD SIDEK,
IR. HIDAYAH BTE BASRI
DR. MOHAMED AHMED HAFEZ

To order additional copies of this book, contact
Toll Free 800 101 2657 (Singapore)
Toll Free 1 800 81 7340 (Malaysia)
www.partridgepublishing.com/singapore
orders.singapore@partridgepublishing.com

ISBN
978-1-5437-5363-9 (sc)
978-1-5437-5365-3 (hc)
978-1-5437-5364-6 (e)

Print information available on the last page.

09/13/2019

ACKNOWLEDGMENT

We are immensely grateful to [TNB] Tenaga Nasional Berhad, and [DID] Department of Irrigation and Drainage Malaysia with exceptional thank you to the Urban Stormwater Management Division for their fully support to this research. We would also like to show our gratitude to ZHL Engineering Consultant Sdn Bhd for sharing their pearls of wisdom with us during this research. We thank our colleagues from [UPM] Dr Aimrun Wayayok and Dr Ahmad Fikri Abdullah who provided insight and expertise that greatly assisted the research.

CONTENTS

CHAPTER I:

INTRODUCTION

1.1 GENERAL INFORMATION

Highland can be defined as an area located at an elevation higher than 500 m above the mean sea level. With slope gradients of more than 25°, these highland areas have the potential to cause landslides and soil erosions if they are not properly managed.

For years, the lands in Cameron Highland have been opened and leveled for agricultural farming and intensive crop production. The overall agricultural coverage is relatively small and is mostly done on steep slopes. The high usage of fertilizer and pesticides by local farmers, accompanied by the increase in the frequency of major storm events had given rise to high levels of soil erosion and environmental pollution

In a study conducted in 2011 (Hamzah et al.), it was discovered that the sustainability of three major agroecosystems in Cameron Highland (namely, tea, vegetables, and floriculture) with respect to soil and nutrient losses and the resulting environmental pollution, has an impact on land resource use and conservation, including the ability to continuously produce hydroelectricity. It is widely known that two of the main economic activities in the Cameron Highlands are hydroelectric power generation and agricultural farming. Most of the present agricultural activities led to serious soil erosion, producing large amounts of sediments. The resulting sediments are then transported and deposited in the storage dam, adversely reducing the hydroelectric power generation capacity and shortening its lifespan.

Inventory data published by the Department of Mineral & Geoscience Malaysia (*Malay: Jabatan Mineral & Geosains Malaysia, JMG*) showed that most of the landslides incidents occurred at areas with intensive agricultural activities, proving that human activities are among the major contributor to soil erosion and landslides occurrences. Therefore, there is an urgent need to minimize and prevent the existence of these incidents in a more holistic manner, which might include both technical and social engagements. As shown in Figure 1.1, based on the data provided by the Department of Mineral & Geoscience, from the total landslides incident recorded in Cameron Highlands (625 nos.), almost 48.16% (301 nos.) occurred in agricultural farming areas, while 51.84% (324 nos.) occurred in non-agricultural related regions. Furthermore, it is estimated that out of the 301 landslide incidents in agricultural areas, 75.42% happened in areas with the sheltered farm, and almost 17.94% occurred in terracing farms. This is shown in Figure 1.2. This analysis is based on data provided by the Department of Mineral & Geoscience and the site inspections conducted at the Cameron Highland areas.

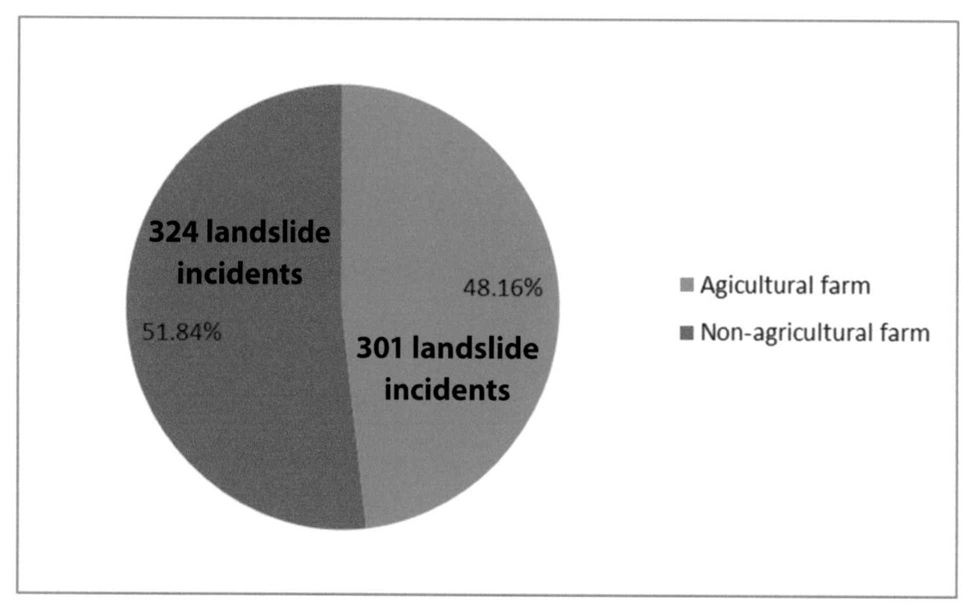

Figure 1.1: Inventory data on Landslide Incidents at Cameron Highlands
(Source: Department of Mineral & Geoscience, 2017)

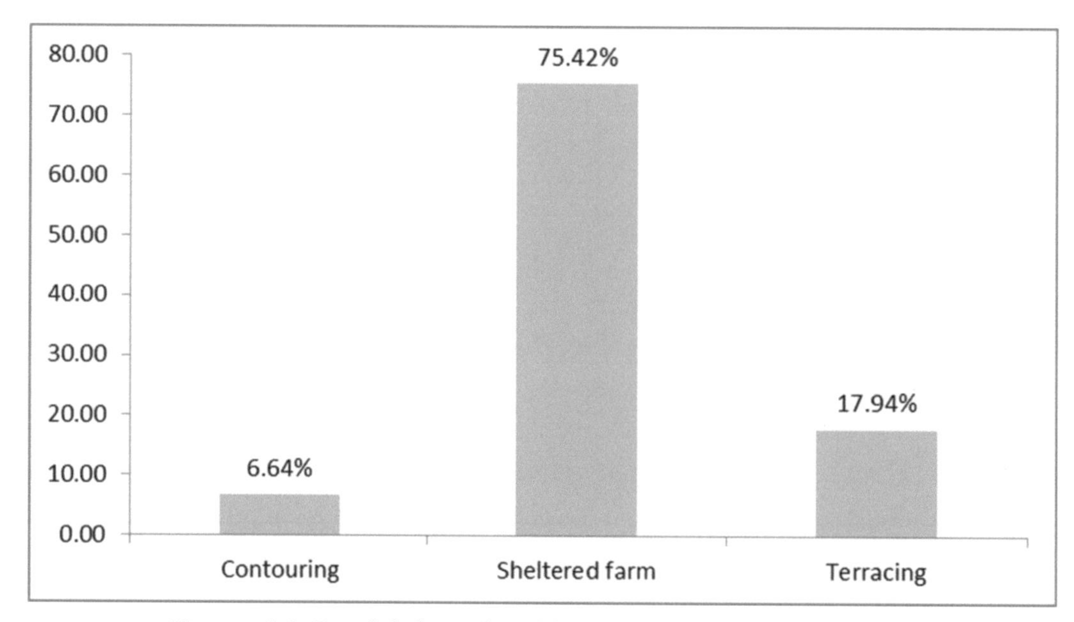

Figure 1.2: Landslide under different Agricultural Practices
(Source: Department of Mineral & Geoscience, 2017)

Cameron Highlands is located on the Main Range, and the surrounding hilly area consists of steep slopes having gradients of more than 20˚ (approximately 59.7% of the surrounding area). Figure 1.3 shows percentages of land area in Cameron Highlands with various terrain classes.

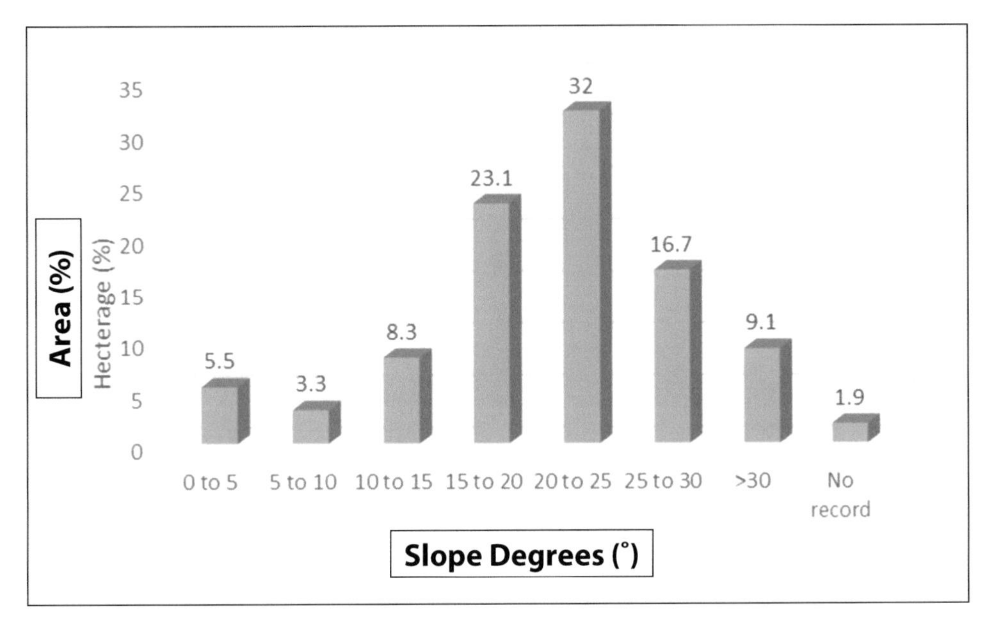

Figure 1.3: Terrain Classes at Cameron Highlands
(Source: Department of Agriculture, 2000)

According to the *Guideline on Agricultural Development on Sloped Areas* published by the Department of Agriculture Malaysia in 2013, the following Table 1.1 shows the Terrain Class in Malaysia:

Table 1.1: Terrain Class in Malaysia

Terrain Class	Steepness		Type of Terrain
	Degrees (°)	Percentage (%)	
C1	0 – 2	0 – 4	*Rata*
C2	2 – 6	4 – 11	*Beralun*
C3	6 – 12	11 – 21	*Berombak*
C4	12 – 20	21 – 36	*Berbukit*
C5	20 – 25	36 – 47	*Sangat Berbukit*
C6	25 – 30	47 – 58	*Curam*
C7	>30	>58	*Sangat Curam*

Source: Department of Agriculture (DOA), 2013

Thus, it can be concluded that only about 17% of the area is below 15°, 23% of the area is between 15° to 20°, and a further 32% is between 20° and 25°. As mentioned before, almost 59.7% of the hillside area in Cameron Highland have a slope gradient of > 20°, which indicates that surrounding areas are hilly to very steep.

Also, the agricultural practices in the highlands comprised of the three major practices; contouring, sheltered and terracing. The slope percentage under different agricultural practices in Cameron Highland is shown in the following Figure 1.4. As can be seen from the figure, the maximum slopes (in percentage) for all three agricultural practices are all in terrain class C6 (*Curam*) while the minimum slopes are mostly in terrain class C3 (*Berombak*). It is worth noting that the DOA 2013 guideline specifies that no agricultural development is allowed at hillside areas with a slope gradient of > 25° (terrain class C6 – *Curam / Steepland*).

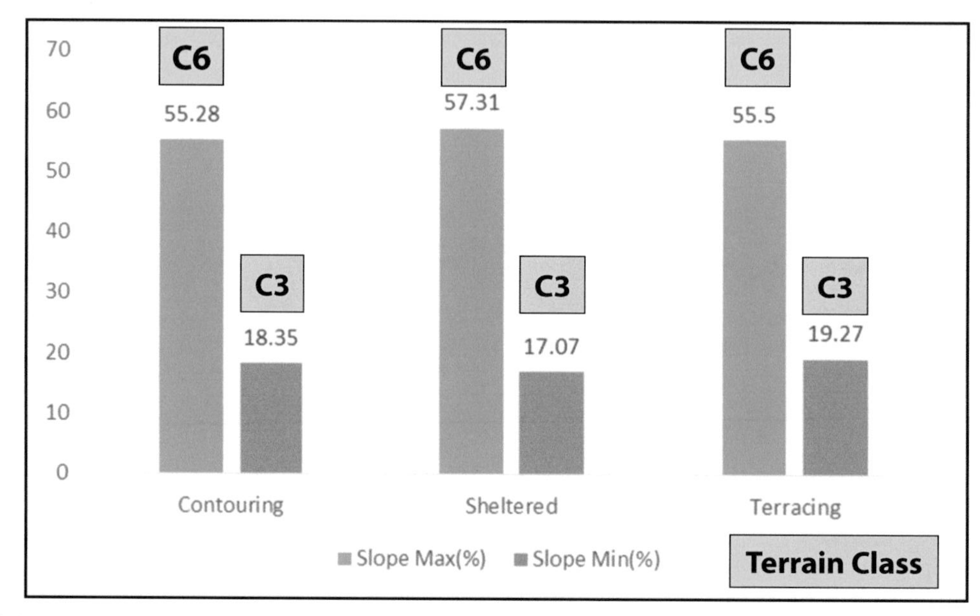

Fig 1.4: Percentage Slopes under Different Agricultural Practices in Cameron Highlands (Source: Department of Agriculture, 2000)

1.2 PURPOSE OF THE GUIDELINE

The authors had developed this Erosion and Sediment Control Guideline for Agricultural Activities in Highland Areas for the following use of:

- To minimize erosion and sedimentation in Cameron Highland, Pahang
- To prevent landslide induced by agricultural farming activities at hilly areas in Cameron Highland, Pahang

This book is to be used by the local authorities and farmers to conserve soil, protect the natural waterways and the surrounding environments from man-made pollutions. It should also be used in conjunction with other published guidelines such as the Guideline for Erosion and Sedimentation Control in Malaysia (DID, 2010) and MSMA 2nd Edition (DID, 2012).

1.3 RULES AND REGULATION GOVERNING THE USE OF LAND AT HILLY AREAS

1.3.1 Soil Conservation Act on Temporary Occupation of Land (TOL)

A. Land conservation act 1960 (part ii - control of hill lands)

This act regulates the planting of short-term crops in hillside lands, prohibiting the cultivation of said crops without a permit unless compliance to the following provisions:

Provided that the Land Administrator may issue an annual permit to plant specified short-term crops to any applicant who satisfies him that such cultivation will not cause appreciable soil erosion, and in such permit may prescribe the area of the land and the terms and conditions under which such cultivation is permitted.

The also laid down restrictions and regulations on the clearing and cultivation of hillside lands:

(1) No person shall clear any hill land or interfere with, destroy or remove any trees, plants, undergrowth, weeds, grass or vegetation on or from any hill land:

Provided that it shall be lawful for the Land Administrator, on the application of the owner or occupier of any hill land, to authorize by permit in writing under his hand, subject to such terms and conditions and such extent and in such manner as may be specified in such permit:

(a) The clearing of such hill land for cultivation;
(b) The clearing or weeding of such hill land under lawful cultivation.

(2) Any person who fails to comply with any terms or conditions prescribed in a permit issued under subsection (1) shall be deemed to have contravened this Act.
(3) Whenever the Land Administrator declines to issue a permit under this section in terms acceptable to the applicant he shall, on being requested so to do by the applicant, forthwith issue to him a certificate under his hand setting forth the nature of the permit asked for and the grounds of such refusal and the date of issue of such certificate.

B. Part III - control of silt and erosion

This part of the act laid down the enforcement provisions (i.e. notice to show cause against order) to be carried out by the Land Administrator. The provisions are as follows:

Whenever it appears to a Land Administrator on grounds to be recorded by him in writing with reference to land owned by any person:

(a) That earth, mud, silt, gravel or stone from such land has caused or is likely to cause damage to other land, whether alienated or not, or to any watercourse, whether natural or artificial, or has interfered or is likely to interfere with the due cultivation of other land, whether alienated or not; or
(b) That by reason of the steepness of the slope of such land, the damage has been or is likely to be caused to such land by erosion or displacement of earth, mud, silt, gravel or stone upon or from such land, the Land Administrator may, by notice served on the owner or occupier of such land, require him to show cause, at a time and place to be stated in such notice, why an order should not be made under this Act prohibiting him from doing, or requiring him to do, any act or thing which may under section 14 be prohibited or required to be done.

This provision also laid down the protocols for the maintenance of work:

Where any drain, watercourse, dam, wall or other work has in pursuance of an order under section 14 been made on any land, all persons who are from time to time registered as owners or occupiers of such land shall, so long as such order remains unrevoked, at his or her own expense maintain such work in good and efficient order to the satisfaction of the Land Administrator.

1.4 USE OF PESTICIDE UNDER PESTICIDE ACT OF DOA

1.4.1 Control of Presence of Pesticide in Lood (Part V): Environmental Quality Act 1974 (Part IV - Prohibition and Control of Pollution)

This act governs restrictions on soil pollution. The provisions are as follows:

(1) No person shall unless licensed, pollute, cause to or permitted to pollute any soil or surface of any land in contravention of the acceptable conditions specified under section 21.
(2) Notwithstanding the generality of subsection (1), a person shall be deemed to pollute any soil or surface of any land if:
  (a) He places in or on any soil or in any place where it may gain access to any soil any matter whether liquid, solid or gaseous; or
  (b) He establishes on any land a refuse dump, garbage tip, soil and rock disposal site, sludge deposit site, waste- injection well or otherwise used land for the disposal of or a repository for solid or liquid wastes so as to be obnoxious or offensive to human beings or interfere with underground water or be detrimental to any beneficial use of the soil or the surface of the land.
(3) Any person who contravenes subsection (1) shall be guilty of an offence and shall be liable to a fine not exceeding one hundred thousand ringgit or to imprisonment for a period not exceeding five years or to both and to a further fine not exceeding one thousand ringgit a day for every day that the offence is continued after a notice by the Director-General requiring him to cease the act specified therein has been served upon him.

It also provides restrictions on the pollution of inland waters:

(1) No person shall unless licensed, emit, discharge or deposit any environmentally hazardous substances, pollutants or wastes into any inland waters in contravention of the acceptable conditions specified under section 21.
(2) Without limiting the generality of subsection (1), a person shall be deemed to emit, discharge or deposit wastes into inland waters if;
  (a) He places any wastes in or on any waters or in a place where it may gain access to any waters;
  (b) He places any waste in a position where it falls, descends, drains, evaporates, is washed, is blown or percolates or is likely to fall, descend, drain, evaporate or be washed, be blown

or percolated into any waters, or knowingly or through his negligence, whether directly or indirectly, causes or permits any wastes to be placed in such a position; or

(c) He causes the temperature of the receiving waters to be raised or lowered by more than the prescribed limits.

(3) Any person who contravenes subsection (1) shall be guilty of an offence and shall be liable to a fine not exceeding one hundred thousand ringgit or to imprisonment for a period not exceeding five years or to both and to a further fine not exceeding one thousand ringgit a day for every day that the offence is continued after a notice by the Director-General requiring him to cease the act specified therein has been served upon him.

1.5 RIVER POLLUTION (EQA AND PAHANG WATER ENACTMENT)

Section 51 of Pahang Water Enactment state that;

(a) It is not allowed to throw a fallen timber into a riverbank,
(b) That Blocking or disturbing river water by wherever means is strictly prohibited
(c) That Construction breaches or heliport is not allowed in the river or beside the river whenever the river size is more than 20 ft.

Section 55 (u) stated that: The authority has the power to enforce laws against sedimentation, pollution, to any rivers, reservoir, catchment or raw water resources

1.6 AGRICULTURAL ACTIVITIES AT CAMERON HIGHLANDS

Agricultural practices in Cameron Highland are divided into three types; terracing, contouring and sheltered farming. Figure 1.5 below shows that almost 58% of farming practices in Cameron Highland can be categorized under sheltered farming, followed by contouring (27%) and terracing (15%).

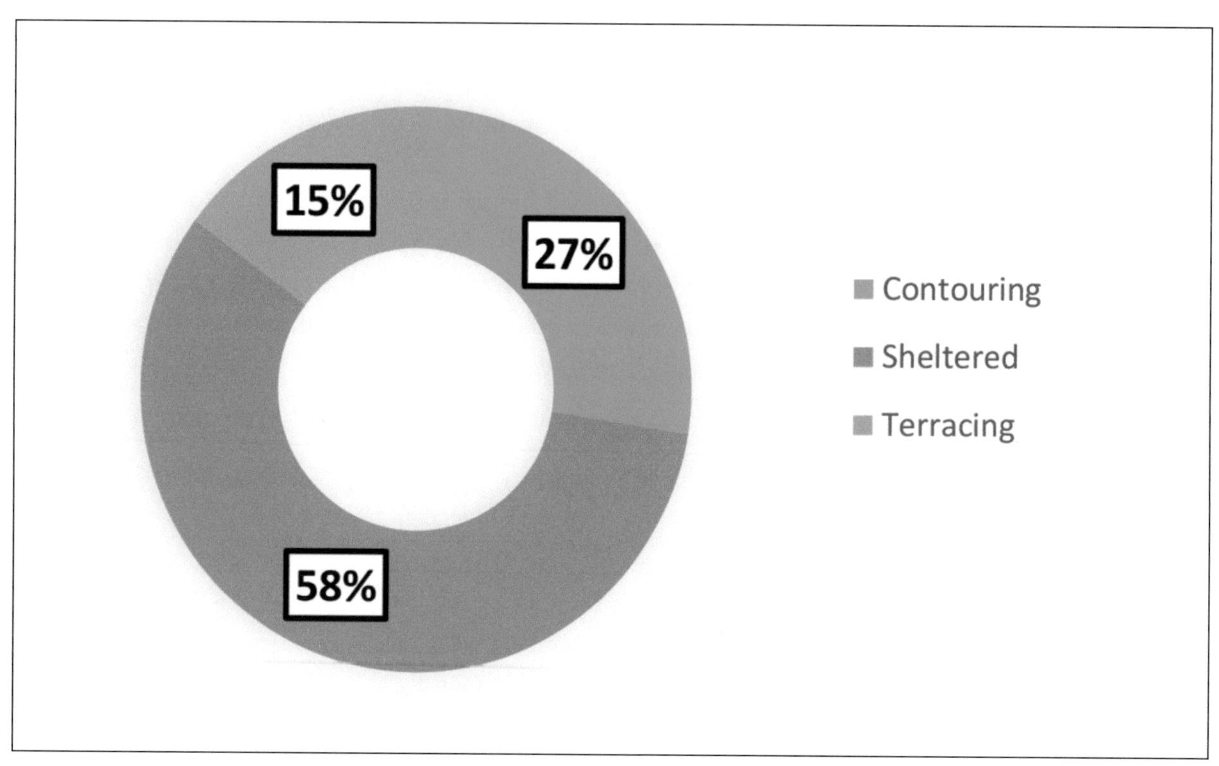

Fig 1.5: Agricultural Practices in Cameron Highlands
(Source: Department of Agriculture, 2016)

Although the topography of the Cameron Highlands is steep and highly dissected, the favourable climate has allowed it to become a major producer of vegetables in Malaysia. Out of the 5,500 ha of agricultural land, vegetables occupy the largest fraction (50%), followed by tea (40%), flowers (7%) and fruits (2%) as reported by Aminuddin et al., (2005). Most of the tea plantations were established in the 1930s by planting tea seedlings on slopes (Figure 1.6). These plantations are managed by large private companies.

It was observed that the tea cultivation is now on the decline due to labour shortages, and in some areas, it is being replaced by vegetables. Three popular vegetable types in Cameron Highlands are cabbage, Chinese cabbage, and tomato. Flower cultivation has increased recently, sometimes the flower is grown at the expense of vegetables, and three major flower species grown in Cameron Highland were chrysanthemum (52%), carnation (20%) and rose (17%). The annual crops are planted on terraces and platforms built on steep slopes or hilltops, as well as on valley floors (as shown in Figure 1.7).

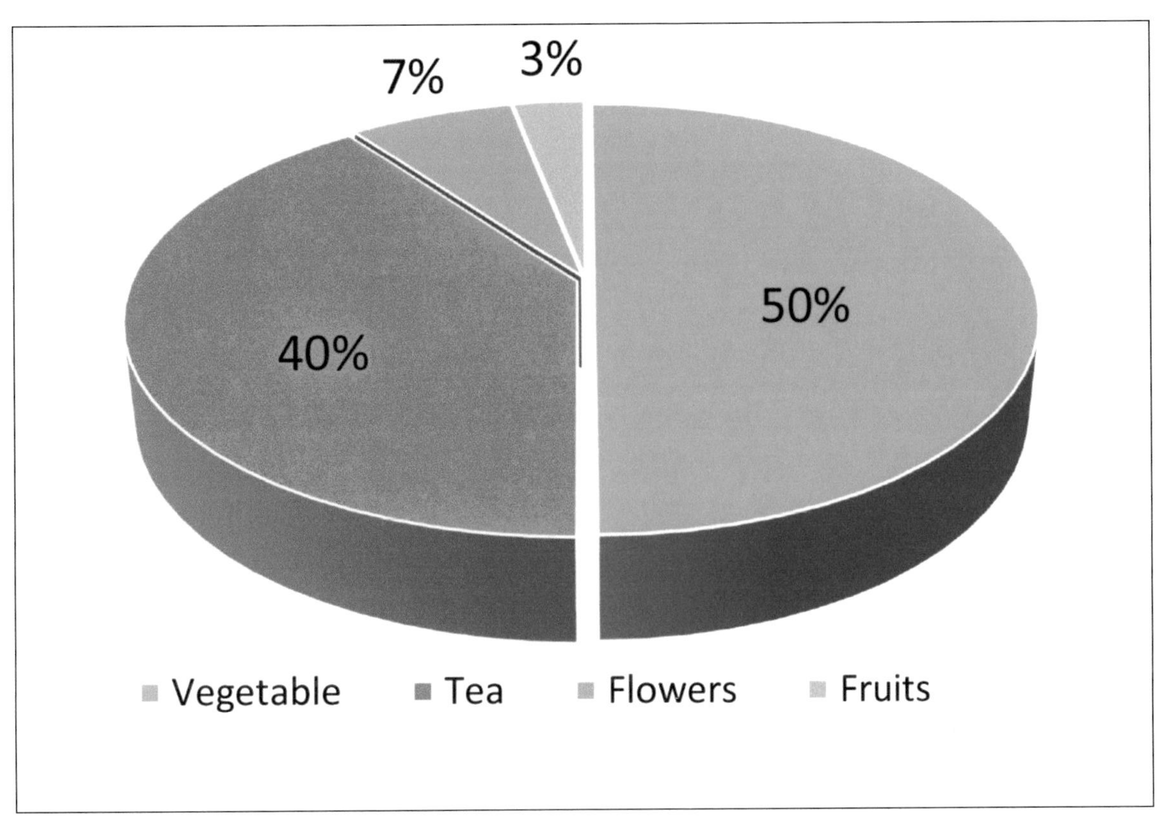

Figure 1.6: Common Cops grew at Cameron Highlands
(Source: Department of Agriculture, 2016)

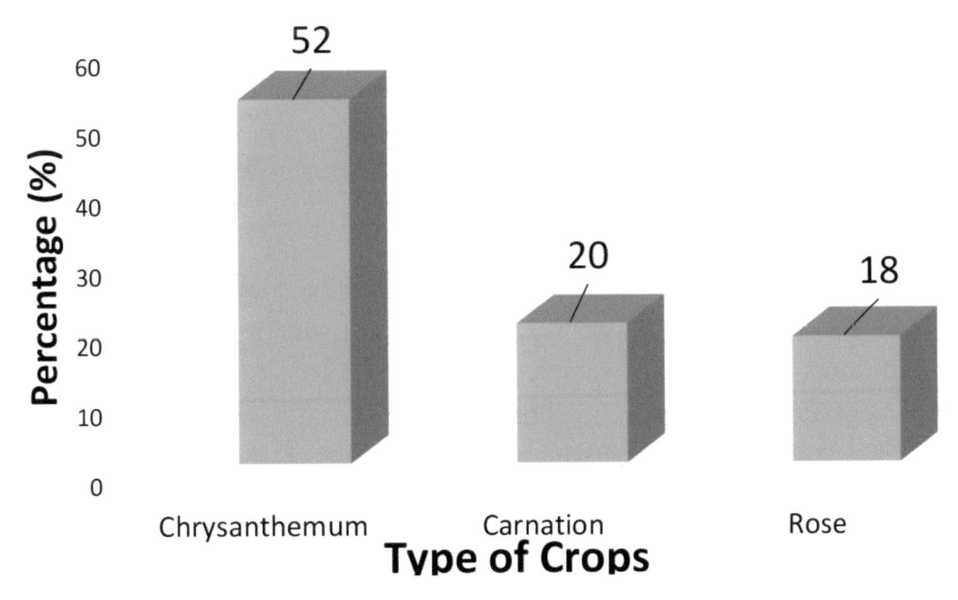

Figure 1.7: Major Flower Species grew in Cameron Highlands
(Source: Department of Agriculture, 2016)

Figure 1.8 below displays the agricultural farming activities near Ringlet, Cameron Highland (Latitude: 04°24'51.44", Longitude: 101°24'18.57"). It can be observed that the annual crops are planted on terraces and platforms built on steep slopes or hilltops, as well on valley floors. Lands are indiscriminately cleared for intensive agricultural production and its related infrastructures. The rate of soil erosion and environmental pollution is expected to increase significantly because the overall agricultural coverage is on steep slopes with high rainfall and high rate of fertilizer and pesticide applications.

Figure 1.8: Existing Agricultural Activities in Cameron Highlands

1.7 OPEN FARM

The procedure of field preparation for crop cultivation in open farms includes deforestation, restumping, and burning. In the case of tea and fruit tree cultivation, a 50% burn usually is sufficient. However, for annual crops like vegetables, complete restumping and burning are essential. In all cases the initial opening of forest results in extensive soil erosion. Tea is planted on slopes, with very little earthwork for its cultivation. A higher level of earthwork is required in the cultivation of fruit trees, which are normally planted on terraces or platforms.

The cultivation of vegetables requires the most extensive earthworks. Vegetables are planted either on terraces or leveled lands depending on the terrain and slope of the land available. There are vegetable plots on valley floors, hill-tops, and on gentle or steep slopes on irregularly shaped hills. In addition, the short cropping cycle of vegetables also poses problems. Immediately after harvest, the fields are prepared for the next crop, and this exposes them to further runoff and erosion. Presently, there are three distinct cultivation seasons in Cameron Highlands for a year with many cycles of land preparations.

Open farming in Cameron Highlands revolves on the ground planting of crops on slope terraces. Tea, for example, is planted on slopes with very little earthwork for its cultivation (Figure 1.9). A higher level of earthwork is required in the cultivation of fruit trees, which are normally planted on terraces or platforms. The cultivation of vegetables requires the most extensive earthworks. Vegetables are planted either on terraces or leveled lands, depending on the terrain and slope of the land available. Also, Figure. 1.10 shows a type of vegetable farming being practice on slope terraces in Cameron Highlands.

In addition, the short cropping cycle of vegetables becomes a problem of great concern because it may result in leaving the bare soil uncultivated. This necessitates the immediate preparation of fields after harvest for the next crop establishment, and this exposes the hilly slope to the risk of further runoff and erosion. Frequently, there are as many as three (3) cultivated seasons a year with many cycles of land preparation.

Figure 1.9: Example of open farming practices (tea plantation) in Cameron Highland

Figure 1.10: Vegetable farming on slope terraces in Cameron Highland

1.8 RAIN-SHELTERED FARMING

Rain-shelters are often used for the cultivation of high-value vegetables and many species of flowers. The land preparation procedure for rain-shelter farming is similar to that for vegetable cultivation, except that for flower cultivation where a plastic rain-shelter is constructed. This is due to the heavy rainfall experienced throughout the year. The plastic roof of the rain-shelters lasts up to 2.5 years. Problems of soil erosion in the flower farms are minimal, as the rain-shelter keeps rainfall out.

However, the volume of runoff intercepted by the shelters is enormous, resulting in sudden large surges of water, sometimes causing flash floods and landslides. Therefore, since the volume of runoff intercepted by the rain-shelters is enormous, well-designed drainage facilities are necessary in order to avoid risk flooding and landslide occurrences. Figure 1.11 to Figure 1.13 showed examples of rain-shelter farming in Cameron Highlands.

Figure 1.11: Cultivation of Vegetables under Rain-shelter

Figure 1.12: Cultivation of flowers inside a rain-shelter

Figure 1.13: Hydroponic farming under a rain-shelter

1.9 FERTILIZER AND PESTICIDE APPLICATIONS

As a result of extensive terracing and land leveling, crops in Cameron Highlands are essentially grown on non-fertile subsoil surfaces. To overcome this problem, fertility is restored through large inputs of both organic and inorganic fertilizers. A high rate of chicken dung application, up to 80 tons/ha, is commonly used in the newly opened areas of Cameron Highlands. Due to heavy rainfall, a major portion of the applied fertilizer is usually lost through runoff and leaching. The heavy input of fertilizers to the soil also results in the soil becoming toxic for plant growth. Farmers, when faced with an unacceptable yield reduction, often add new soil material over the existing soil or simply obtain fresh topsoil. Cultivated vegetables and flowers are susceptible to a number of pests and diseases. Chemical control remains the most popular approach to pest and disease management. Although integrated pest management (IPM) strategies are being promoted by the relevant government agencies, the use of pesticides is still significant. The usage of organo-phosphorus and synthetic pyrethroid compounds is widespread among vegetable farmers (Aminuddin et al., 2005). Herbicides such as glyphosate and paraquat are commonly applied for weed clearing purposes before the crop establishment.

1.10 EFFECT OF SOIL EROSION

1.10.1 Effect On-Site

The implications of soil erosion by water extend beyond the removal of valuable topsoil. Crop emergence, growth, and yield are also directly affected by the loss of natural nutrients and applied fertilizers. Seeds and plants can be disturbed or entirely removed by the erosion.

Organic matter from the soil, residues and any applied manure is relatively lightweight and can be readily transported off the field, particularly during high rainfall intensity conditions. Pesticides may also be carried off the site with the eroded soil. Soil quality, structure, stability, and texture can be affected by the loss of soil. The breakdown of aggregates and the removal of smaller particles or entire layers of soil or organic matter can weaken the structure and even change the texture. Textural changes can, in turn, affect the water-holding capacity of the soil, making it more susceptible to extreme conditions such as drought. Finally, erosion leads to a reduction of agricultural productivity due to lacking soil-water availability and the less fertile soil, which was eroded.

1.10.2 Effect Off-Site

The off-site impacts of soil erosion by water are not always as apparent as the on-site effects. Eroded soil, deposited downslope, inhibits or delays the emergence of seeds, buries small seedlings and necessitates replanting in the affected areas. Also, sediment can accumulate on down-slope properties and contribute to road damage. Sediment that reaches streams or watercourses can accelerate bank erosion, obstruct stream and drainage channels, fill in reservoirs, damage fish habitat and degrade downstream water quality. Pesticides and fertilizers, frequently transported along with the eroding soil, contaminate or pollute downstream water sources, wetlands and lakes. Because of the potential seriousness of some of the off-site impacts, the control of "non-point" pollution from agricultural land is an important consideration. Soil erosion causes loss of surface soil layers containing organic and mineral nutrient pools, partial or complete loss of soil horizons and possible exposure of growth-limiting subsoil, as well as off-site impacts such as damage to private and public infrastructure, reduced water quality, and sedimentation.

1.11 EROSION RATE CLASSIFICATION

From the soil erosion calculation for Cameron Highlands, the two catchments covered were Telom and Bertam Catchments. From the analysis, the soil loss by erosion from the catchments was respectively 38.0 t ha^{-1} year^{1} and 73.9 t ha^{-1} year^{1}. Comparing with erosion classification in Table 1.2, the soil erosion in Cameron Highlands is at the high side.

Table 1.2: Severity Level of Soil Erosion

Level	Soil Loss (t ha^{-1} year^{-1})
Very Low	0-1
Low	1-5
Medium	5-15
High	>15

Source: Department of Agriculture (DOA)

CHAPTER II:

DESIGN GUIDELINES FOR EROSION & SEDIMENT CONTROL BMPS FOR AGRICULTURAL FARM

2.1 SOIL LOSS AND SEDIMENT YIELD ESTIMATION

Modified Universal Soil Loss Equation (MUSLE) is relevant and encouraged to use for sediment yield estimation of a catchment as a result of a particular rainfall event (Williams, 1975). The estimated amount of sediment storage volume is used in the sediment basin/ trap design. This empirical relationship is elaborated and expressed clearly (refer to MSMA 2nd edition, 2012).

2.1.1 Soil Loss Estimation Using the Universal Soil Loss Equation (USLE)

The annual average erosion from a given geographical area and time could be computed by using the Universal Soil Loss Equation (USLE) (Equation 2.1). USLE is an erosion model developed by the United States of America, Department of Agriculture to support the decision in soil conservation planning and management. The model was further modified to improve the prediction accuracy and to be applied universally.

Smith et al. (1947) presented a method for estimating soil losses from fields of claypan soils. Soil loss ratios at different slopes were given for contour farming, strip cropping, and terracing. Recommended limits for slope length were presented for contour farming.

The USLE computes the average annual soil loss as follows:

$$A = R.K.LS.C.P \qquad (2.1)$$

Where:

A = annual soil loss (tones/ha/year)

R = rainfall factor (MJ/mm/ha/ha/yr)

K = soil erodibility factor (ton hour/ MJ/mm)

L x S = slope length and steepness factors, respectively (dimensionless)

C = vegetation and management factor (dimensionless)

P = support practice factor (dimensionless)

The Modified Universal Soil Loss Equation (MUSLE) is perhaps the most frequently used equation (Equation 2. 2) for sediment yield estimation. It is developed by Williams (1975) to calculate sediment yields of a catchment as a result of a specific storm event. This empirical relationship is expressed by the following equation for individual storm events.

$$Y = 89.6\ (VQp)^{0.56}\ (K.LS.C.P) \quad (2.2)$$

Where:

Y = Sediment yield per storm event (tonnes)

V = Runoff volume (m^3)

Q_p = Peak Discharge (m^3/s)

K, LS, C, P = USLE Factors

For further details, please refer to MSMA (2012) 2nd Edition

2.2 SOIL LOSS ESTIMATION USING GEOSPATIAL METHOD

This method utilized geospatial tools (such as geospatial information system and remote sensing) to estimate the amount of soil loss for a certain study area. Figure 2.1 shows the spatial image for landslide susceptibility in Cameron Highland developed using geospatial methods. Figure 2.2 and 2.3 meanwhile, shows the Land Cover Factor (C) developed for Cameron Highland and the Conservation Practice Factor (P) in Cameron Highland. These new values for C and P are to be used in conjunction with the existing Cover Management (C) and Support Practice (P) Factor value established in MSMA Chapter 2 (as shown in Table 2.1 – Table 2.4). Furthermore, Figure 2.4 shows the erosion modelling procedures.

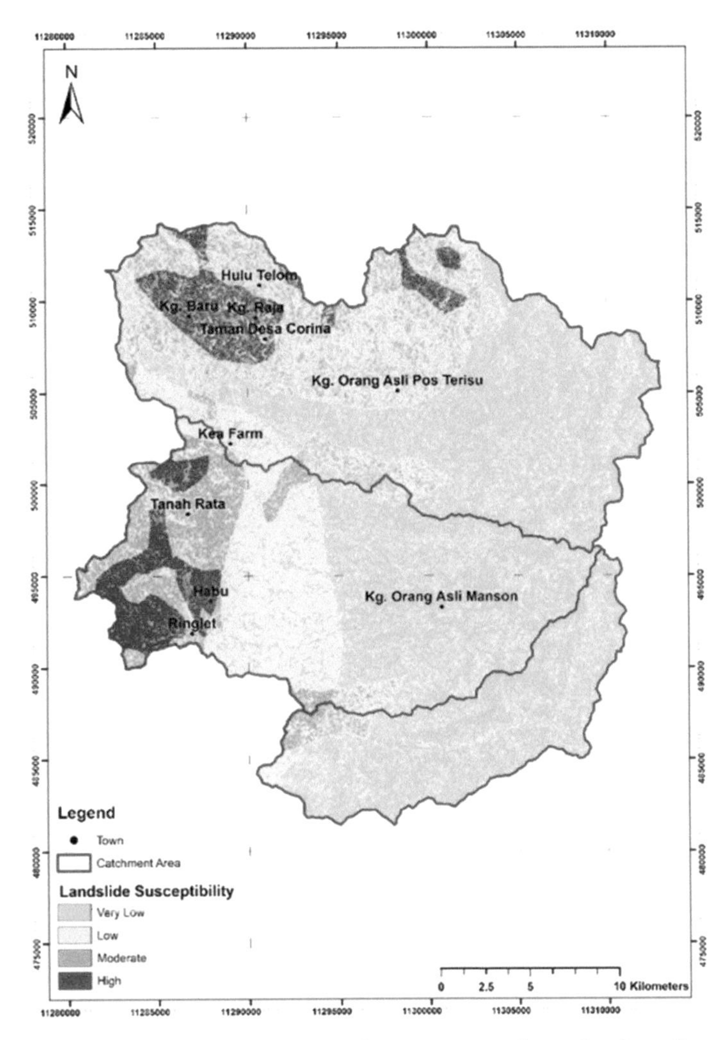

Figure 2.1: Landslide susceptibility using the geospatial method at Cameron Highlands

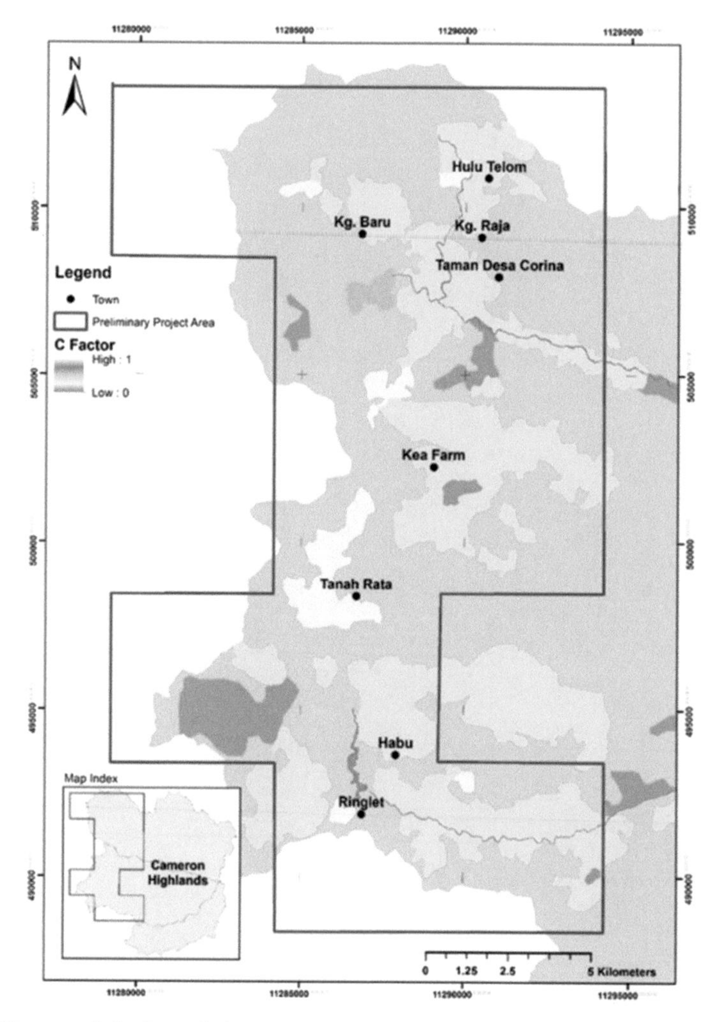

Figure 2.2: Land Cover Factor (C) of Cameron Highlands

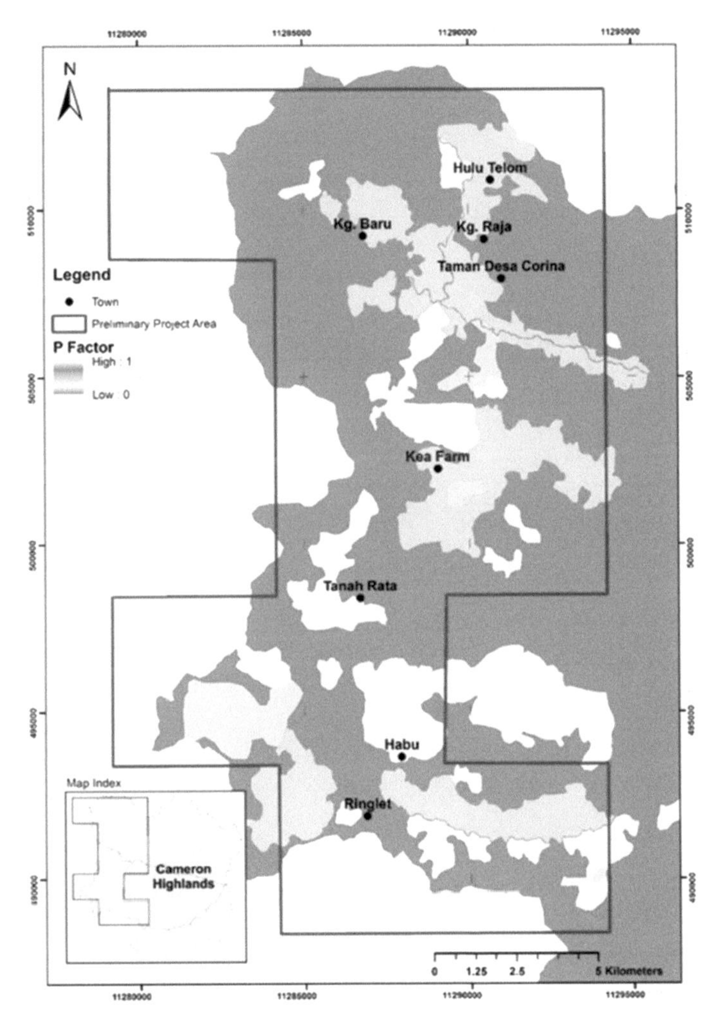

Figure 2.3: Conservation Practice Factor (P) of Cameron Highlands

Table 2.1: Cover Management, C Factor for forested & undisturbed lands

Erosion Control Treatment	C Factor
Rangeland	0.23
Forest / Tree	
25% Cover	0.42
50% Cover	0.39
75% Cover	0.36
100% Cover	0.03
Bushes / Scrubs	
25% Cover	0.40
50% Cover	0.35
75% Cover	0.30
100% Cover	0.03
Grassland (100% coverage)	0.03
Swamps / mangrove	0.01
Waterbody	0.01

Source: MSMA 2nd Edition, DID, 2012

Table 2.2: Cover Management, C Factor for agricultural & urbanized areas

Erosion Control Treatment	C Factor
Mining Areas	1.00
Agricultural Areas	
Agricultural Crops	0.38
Horticulture	0.25
Cocoa	0.20
Coconut	0.20
Oil Palm	0.20
Rubber	0.20
Paddy (with water)	0.01
Urbanized Areas	
Residential	
Low Density (50% Green Area)	0.25
Medium Density (25% Green Area)	0.15
High Density (5 % Green Area)	0.05
Commercial, Educational & Industrial	
Low Density (50% Green Area)	0.25
Medium Density (25% Green Area)	0.15
High Density (5 % Green Area)	0.05
Impervious (Parking Lot, etc.)	0.01

Source: MSMA 2nd Edition, DID, 2012

Table 2.3: Cover Management, C Factor for BMPs at construction sites

Erosion Control Treatment	**C Factor**
Bare soil / Newly cleared land	1.00
Cut and fill at construction sites	
Fill	
Packed, smooth	1.00
Freshly disked	0.95
Rough (offset disk)	0.85
Cut	
Below root zone	0.80
Mulch	
Plant fibers, stockpiled native materials / chipped	
50% cover	0.25
75% cover	0.13
100% cover	0.02
Grass-seeding and sod	
40% cover	0.10
60% cover	0.05
>90% cover	0.02
Compacted gravel layer	0.05
Geo-cell	0.05
Rolled Erosion Control Product:	
Erosion Control Blankets / Turf Reinforcement Mats	0.02
Plastic Sheeting	0.02
Turf Reinforcement Mats	0.02

Source: MSMA 2nd Edition, DID, 2012

Table 2.4: Support Practice, P Factor for BMPs at construction and development sites

Erosion Control Treatment	**C Factor**
Bare soil	1.00
Disked bare soil (rough or irregular surface)	0.90
Wire log / Sand bag barriers	0.85
Check Dam	0.80
Grass buffer strips (to filter sediment-laden sheet flow)	
Basin slope (%)	
0 to 10	0.60
11 to 24	0.80
Contour furrowed surface	
(maximum length refers to downslope length)	
Slope (%) Max Length (m)	

Erosion Control Treatment		C Factor
1 to 2	120	0.60
3 to 5	90	0.50
6 to 8	60	0.50
9 to 12	40	0.60
13 to 16	25	0.70
17 to 20	20	0.80
>20	15	0.80
Silt fence		0.55
Sediment containment systems		0.50
Berm drain and cascade		0.50
Terracing		
Slope (%)		
1 to 2		0.12
3 to 8		0.10
9 to 12		0.12
13 to 16		0.14
17 to 20		0.16
>20		0.18

Source: MSMA 2nd Edition, DID, 2012

Description
• Conservation practices factor (P) explains the effect of practices concerning sheltered crop, unsheltered crop, terraces, silt fences, subsurface drainage, etc. • In some large-scale regions, alteration of support practices cannot be diagnosed via a land-use map only. Therefore, field investigation and data collection must be done to gather the information required. • At this stage, the information on the sheltered agricultural farming area had been digitized from the 2m resolution photo of the project area. From the observation and analysis using ArcGIS, it is found out that 57.6% of the agricultural in the project area was sheltered.

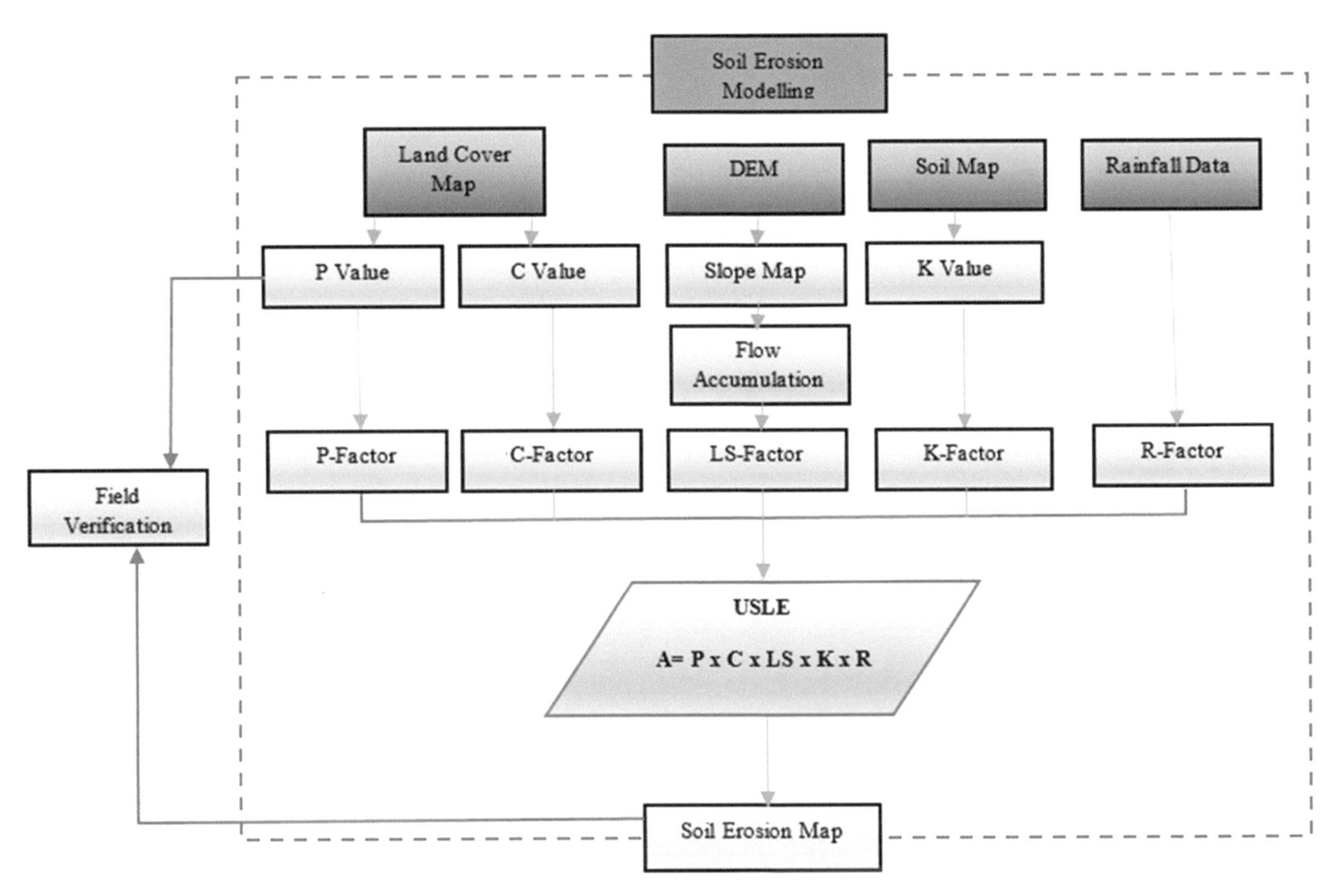

Figure 2.4: Soil erosion susceptibility model development chart

Details of soil erosion and landslides susceptibility maps at Cameron Highlands are presented in Appendix E of this guideline.

2.3 SLOPE CALCULATIONS FOR OPENING NEW FARM IN HILLY AREAS

Management of hillside areas would require the modifications of existing equations for areas with steep and long slopes which often generates a high amount of runoffs at high velocity. This large amount of runoff can carry the soil particles away from their original positions.

Cameron highland is one of those areas in Malaysia that are characterized with very steep slope gradients, thereby generating a high amount of runoffs that are often the cause of soil erosion. The sediments resulting from this are usually deposited into conveyance structures such as streams and/or reservoirs, causing massive siltation problems for the surrounding rivers and dam reservoirs.

2.4 SOIL STRESSES STABILITY COMPUTATION

Soil normal stress (σ) is defined by the ratio of force acting normal to the surface (F_n) to the area perpendicular to the direction of force (A). When the reasonable force increases, it increases the cases of failure (equation 2.3).

The shear stress (τ) refers to the ratio of tangential force on acting on soil parallel to the surface. This shear stress increases with the increase of normal stress (equation 2.4).

$$\text{Normal} = F_n / A = \sigma \quad (2.3)$$
$$\text{Shear} = F_t / A = \tau \quad (2.4)$$

Where:

F_n = Normal force (N)

F_t = Tangential or shear force (N)

As $\sigma \uparrow \tau$ to cause failure $= \tau_f \uparrow$

$\tan \Phi = (\tau_f / \sigma)$; Φ = angle of internal friction

2.5 SOIL STRENGTH AND MECHANISM

Each soil type has its own different Φ value. Table 2.5 and Figure 2.5 below displays the range of soil types with their corresponding Φ values. Also, Figure 2.6 displays the different types of slope failures. The value of Φ (ultimate or peak) selected for use in practical soil or foundation problems should be related to the soil strains that are expected. If soil deformation will be limited, using the peak value for Φ would be adjusted. Where deformations might be relatively large, the ultimate values of Φ should be used.

Table 2.5: Representative of Φ Values For non-cohesion Soils

Soil type	**Angle Φ, in Degrees**	
	Ultimate	**Peak**
Sand and gravel mixture	33 – 36	40 – 50
Well-graded sand	32 – fao35	40 – 50
Fine to medium sand	29 – 32	32 – 35
Silty Sand	27 – 32	30 – 35
Silty (non-plastic)	26 – 30	30 – 35

Source: adopted from Raj (2002)

2.5.1 Footing Bearing Loads

The footing bearing of a slope can be determined from the following relation (equation 2.5)

$$q_{ult} = a_1 * c * N_c + a_2 * B * \gamma_1 * N_\gamma + \gamma_2 * D_f * N_q \quad (2.5)$$

Where:

c = soil cohesion beneath footer

γ_1', γ_2 = effective soil unit weight above and below footer

B = footer size term

N_c, N_γ, N_q = capacity factors

D_f = footing depth below surface

Therefore,

$$q_{design} = q_{ult} / FS \quad (2.6)$$

Table 2.6: The values of B, a1 and a2 for different slope sizes

Length/width	B	a_1	a_2
1 (square)	Width	1.2	0.42
2 (square)	Width	1.12	0.45
3 (square)	Width	1.07	0.46
4 (square)	Width	1.05	0.47
6 (square)	Width	1.03	0.48
Strip	Width	1.00	0.50
Circular	Radius	1.2	0.60

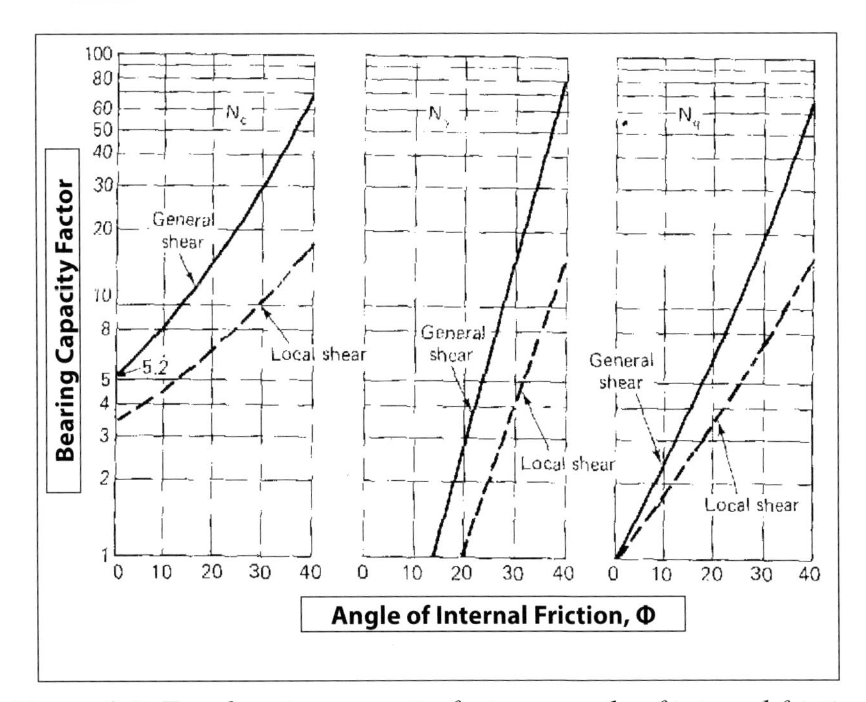

Figure 2.5: Foot bearing capacity factor vs angle of internal friction

2.6 SLOPE STABILITY AND FAILURE

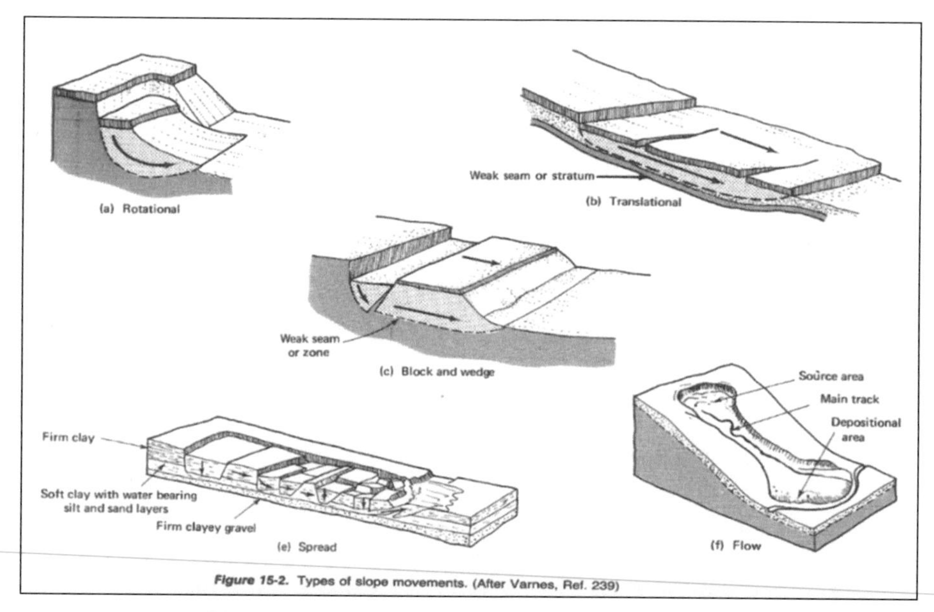

Figure 2.6: Possible forms of Slope failure

The terms of consideration while accessing slope failures are;

β = max. slope angle before sliding

Φ = angle of internal friction

The condition of Cohesion less soil;

$\tan(\beta) = \tan(\Phi)$ and the ratio of;

Saturated: $\tan(\beta) = (1/2)\tan(\Phi)$

The condition for cohesive soil;

$\gamma * z * \sin(\beta) * \cos(\beta) = c + \sigma * \tan(\Phi)$

Where;

z = assumed depth

c = cohesive force

σ = effective compressive stress rotational or sliding block

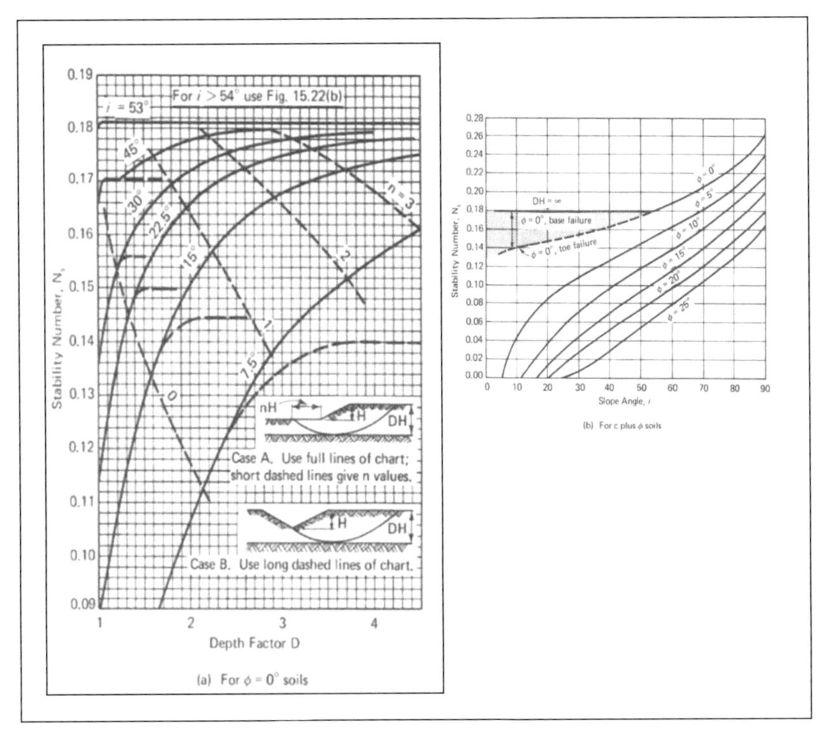

Figure 2.7: Slope stability curves
(*Source: McCathay, 1982***)**

To calculate the Maximum height of slope (H_{max}), the following equation 2.7 is used.

$$N_s = c / (\gamma * H_{max}) \qquad (2.7)$$

Where,

c = cohesion force

γ = soil unit weight

H_{max} = max depth without sliding

2.7 EROSION CONTROL BMPS FOR OPEN FARM

Agricultural activities in hilly areas are exposed to the risk of soil erosion. Soil erosion will cause the top layer of the fertile soil to be washed away by rainfall and surface runoffs, thereby depleting the area of its essential soil nutrients. Furthermore, due to high-velocity surface runoff, water would not have sufficient time to be absorbed by the soil, thereby affecting the groundwater. In conclusion, crops in those areas suffer both the loss of soil fertility and groundwater. Therefore, it is necessary to introduce erosion control measures, not just to prevent environmental pollution at the downstream water body, but also for the benefits in crop production. Figure 2.8 shows the soil erosion measure chart for agricultural farming.

In summary, the approach to BMPs soil erosion control strategies are based on the following:

i. Covering the soil to protect it from raindrop impact.
ii. Increase soil infiltration capacity to reduce runoff
iii. Improve aggregates stability of the soil
iv. Increasing surface roughness to reduce runoff velocity.

The soil conservation techniques/measures can be grouped into:

(a) Agronomic and soil management measures
(b) Engineering (mechanical) measures
(c) Rain Water Harvesting (RWH) For Agricultural Areas

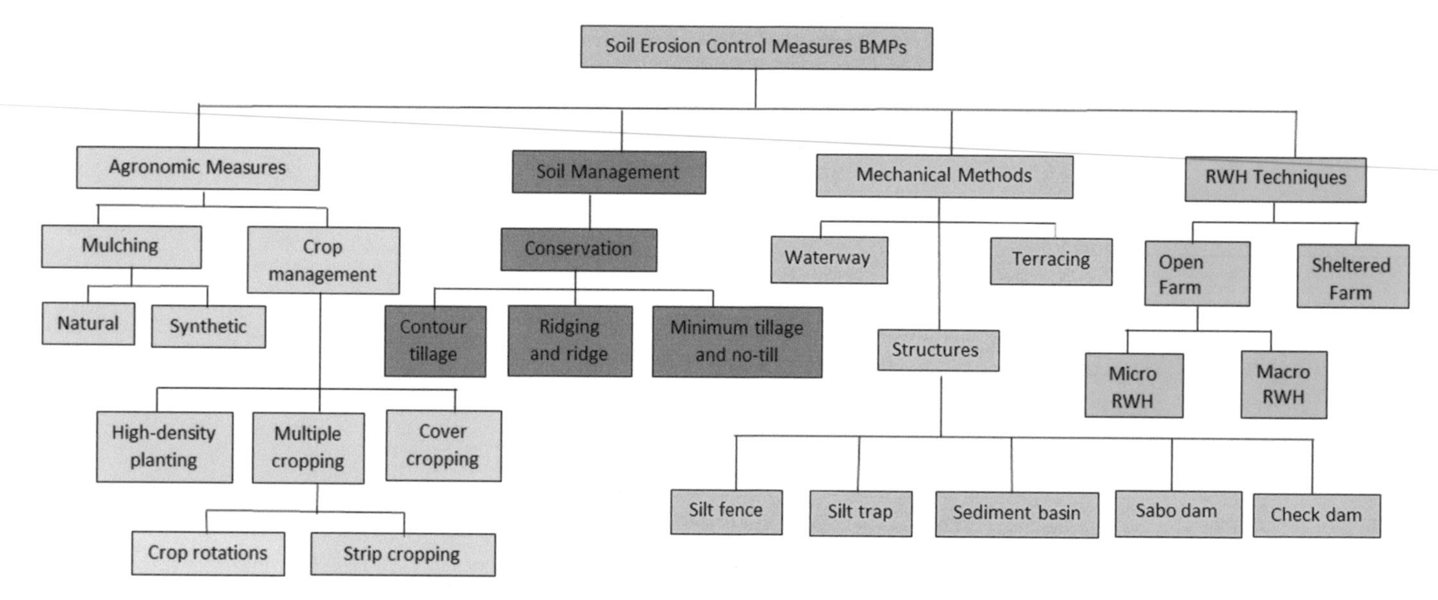

Figure 2.8: Soil Erosion Measures (BMPs) for agricultural farming

2.8 AGRONOMIC MEASURES BMPS FOR EROSION CONTROL FOR OPEN FARMS

Agronomic measures refer to the role of vegetation to minimize soil erosion. These measures offer protection to the soil by covering the soil surface, increasing surface roughness, increasing surface depression storage and increasing soil infiltration capacity. For BMPs control measures in unsheltered farms, the agronomic activities should be done in such a way that the minimum soil disturbance is achieved with least soil erosion occurrences. This consists of the following practices;

i. Contour cultivation
ii. Cropping systems
iii. Tillage practices
iv. Soil management practices

2.9 CONTOUR CULTIVATION FOR BMPS

The BMPs contour cultivation involves carrying out agricultural operations such as plowing, planting, cultivating, and harvesting on the contours with minimum soil erosion. Since the contour cultivation is to be done in open farms, this reduces surface runoff by impounding water in a small depression and decreases the development of rills. In some cases, after the inter-cultivation operations, a ridge and furrow system on the contour can offer greater resistance to surface runoff. To layout the system on the field, the guide should be marked across the slope using dumpy or hand level. All the subsequent agricultural operations are carried out making reference to the marked lines. Contouring on steep slopes or under high rainfall intensity condition, soil erosion ability is low, and this will decrease the rate of gully formation.

Description
• The first step is to determine the contour line at the field • Crops are then planted across the slope or "on the contour". This allows the farmers to do all the land preparation, planting, and harvesting across the slope. • Farming across the slope helps to shorten slope lengths, slowing down runoff water so it can be soaked into the soil. • It is important to take time to lay out the contour lines to guide the farmers in the field. This is done with simple instruments such as a level or a homemade A-frame. • The local office of cooperative extension service can offer help to show the farmers how to lie out and race contour lines.

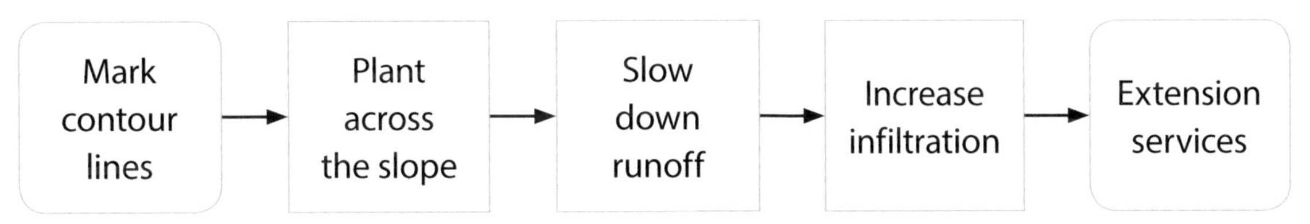

Figure 2.9: Procedure for contour cultivation

2.9.1 *Cropping Systems for BMPs*

Cropping system refers to a sequence of crops growing on a given area over some time. The cropping system should be designed to achieve BMPs control measures if it achieves the following: maintain soil fertility, protect the soil from erosion and making the best use of available soil moisture. Cropping systems includes: Strip cropping, Crop rotation, Intercropping/mixed cropping.

A. Strip Cropping

Strip cropping is the practice of growing alternate strips of different crops in the same field to serve as vegetative barriers to erosion. For controlling water erosion, the strips are on the contour. The alternate strips should consist of close-growing erosion-resistant crops to erosion permitting crops. The rotational arrangement should be such that erosion-resistant crops follow erosion allowing crops. To achieve the best results, strip cropping is to be done in combination with other farming practices, such as crop rotations, contour cultivation with good land management practices. Tillage and field layout is held closely to the contour, and the crops follow a definite rotational sequence.

Table 2.7: Criteria for strip cropping

Items	Description / Criteria
Contour line Installation method	• The first step is to determine the alignment of the contour line at the field • Establish a point for locating the contour line that will form the lower boundary of the first strip. • This point is located at a predetermined width (say, 25m) apart from the top boundary of the field by measuring along the steepest part of the stop. • A contour line is then drawn passing through this point up to the field boundary. • This procedure is repeated until the entire field is laid out.

The width of Strips varies with the degree and length of the land slope, allowable soil loss, soil types, arrangements of crops grown in rotation and size of farm equipment used in the field. The width of the strip is adjusted according to the terrace interval, but in untraced areas, narrow width than the standard terrace interval is frequently used. In general steeper the slope, the narrower will be the strips for BMPs of cultivated and dense-growing crops both. An approximate range of trip widths based on average land slope and soil types is given in the table below:

Table 2.8: Soil type and corresponding strip width

Soil Types	Strip width (m)
Sandy soil	6.0
Loamy Sand	7.0
Sandy Loam	30.0
Loam	75.0
Silt Loam	85.0
Clay Loam	105.0

Source: adapted from Troeh (2013)

The widths of the strip may vary depending on the slope of the field, soil texture, rainfall characteristics, type of crops and equipment availability. In general, the steeper the slope, the greater the width of the erosion resisting crop and the smaller the width of erosion permitting crop for BMPs. Also, a ratio of 5:1 is recommended for BMPs. There exist four types of strip cropping. These are contour strip cropping, field strip cropping, buffer strip cropping, and wind strip cropping (Figure. 2.10).

Additional criteria:

- Infield strip cropping, the strips are laid across the slope in uniform width without taking into consideration the exact contours.
- This method can be practised in the field with regular slopes and soil of high infiltration rate.
- When used with adequate grass drainage, the strips may be placed where the topography is too irregular to make contour strip cropping practical.

- In buffer strip cropping, permanent strips of grasses are located either in severely eroded areas or in areas that do not fit into regular rotation.
- Buffers may be even or irregular in width or placed on critical slopes areas of the field.
- The type used depends on cropping system, topography, and types of erosion hazard, a buffer strip is more or less in a permanent contour usually varies in width which is usually kept between 3 to 5 m for BMPs.

Figure 2.10: A typical field with strip farming in Cameron Highlands

B. Crop Rotation for BMPs

Crop rotation is used purposely to prevent or control soil erosion, ensure vegetative cover and build-up soil fertility. Crop rotations can be more useful for controlling soil erosion accompanied by strip cropping system. BMPs are achieved when legumes are included in the rotation which helps to build up organic matter content and to improve the soil's physical conditions. The suitability of crop rotation often differs from one region to another, depending on the type of soils and climate. For Cameron Highland, crop rotation is highly recommendable.

Table 2.9: Criteria for crop rotation

Items	Description / Criteria
Location	• Can be utilized on the same piece of land by growing tilled crops; small grain crops, hay crops or grasses either under a strip cropping system or a separate field system. In areas where perennial grasses and legumes are not feasible to grow, the row crops of a small grain and annual legume crops can also be grown in strips. It is a general rule that no two cultivated strips should have the same planting or harvesting dates.
Sequence of crops	• The sequence of crops should be in such a manner that there could form a dense - fibrous root system to hold the soil and retard the erosion until the roots are broken down by tillage operations.

As mentioned, crop rotation also increases the organic matter in the soil thereby the physical condition of the soil become improved ultimately soil absorbs more water and also increases the capability of soil to resist the erosion. Underuse of crop rotation practices for controlling soil erosion, the simplest way to combine different crops in rotation form and grow them in consecutive rotations.

The frequency with which row crops should be grown depends upon the severity of erosion taking place in the area. For example, where erosion rate is very low the row crops can be produced at every alternate year, but on the contrast in high erodible areas or where soil erosion is being more intense like Cameron Highlands, there may be practised only once in five or even seven years cycle for BMPs.

For BMPs erosion control by growing the crops in the notation system, probably the most suitable crops are legumes and grasses. The main benefits of these crops are;

a) Reduction of soil erosion resulting from a high degree of good ground cover.
b) Help to maintain or improve the status of organic content in the soil, thereby contributing the soil fertility and enable to develop more stable aggregates in the soil.
c) Increase in soil nitrogen resulting from nitrogen fixation associated with legume crops.

C. Mixed cropping

Mixed cropping is the system of growing more than one crop together in the same field. Different crops may be sown in separate rows according to a pattern in which some cases referred to as intercropping. Erosion control can be achieved by growing erosion resisting and erosion permitting crops, growing tall and short crops to reduce wind effect for BMPs.

Table 2.10: Criteria for mixed cropping system

Items	Description / Criteria
Crop mixture	• There should be a mix of deep-rooted and shallow-rooted crops. Similarly, a combination of early maturing and late maturing crops, wide-spaced and close-spaced crop, legume and non-legume crops, erosion permitting and erosion resisting crops are also recommended. • On occasions, selections are made based on leaf density and area index. This is because soil coverage is important to ensure protection against rainfall and thus, causing less erosion

Fig 2.11: Mixed cropping system / high-density farming
(Source: Farming Programs on the Eastern Seaboard, 2017)

Some erosion control guidelines stated that the most cost-effective measure that could minimize the volume of contaminants such as sediments and pesticides from discharging into the nearest streams or rivers through overland flow is the planting of cover crops (*BMPs, 2013*).

Protection of bare soil surfaces is one of the best ways to prevent soil loss. Depending on the type, grasses provide short-term and long-term soil stabilization for disturbed areas. The crops roots effectively hold the soil together in place, thereby preventing the displacement of soil due to seepage and surface runoff. Long-term use of the cover crops also provides an added benefit of increasing water infiltration into the sub-soil, therefore, reduces the surface runoff that often causes surface soil erosion.

In addition to its ability of sheet erosion prevention, cover crops could also serve many other agronomic purposes such as fixing nitrogen in the soil and providing habitat for beneficial insects.

D. Types of Surface Soil Cover

a) Cover Crops

In practice, various types of plants and vegetation are in used as cover crops. According to the handbook published by the Sustainable Agricultural Research & Education (SARE) in 2007 titled, *Managing Cover Crops Profitably (3rd Edition)*, the types of cover crops can generally be classified into two categories as:

i. Legume Cover Crops
ii. Non-Legume Cover Crops

Legume is a plant originating from the family, *Leguminosae*, which is one of the most common species found in the tropical rainforest around the world, in addition to the dry forest in the America and African regions. There are many types of legumes, most of which produce a pod that splits evenly in into two (peas and beans).

Legumes are used for;

- Nitrogen fixation (the biological process of changing atmospheric nitrogen gas into ammonia using bacteria) that generates nitrogen in the soil for the benefit of surrounding plants,
- Reduce or prevent erosion,
- Attract beneficial insects,
- Produce biomass and add organic matter to the soil as natural fertilizers.

It must be noted that legumes vary widely in their ability to prevent erosion, suppress weeds and add organic matter into the soil. The most commonly used legume cover crops used in the United States are berseem clover, cowpeas, crimson clover, field peas, hairy vetch, medics (*Black Medic*), red clover, subterranean clovers, sweet clovers, white clover, and woolly pod vetch.

It must also be noted that there are many factors that might affect the effectiveness of each cover crops, ranging from regional soil properties to seasonal changes. For instance, winter-annual legumes are usually planted in autumn and often produced most of their biomass and available nitrogen (N) in the spring. Thus, depending on the climate, management of legume cover crops in the spring would often involve balancing the early planting of the cash crop with waiting to allow more biomass and nitrogen (N) to be produced by the legume. Therefore, the establishment and management of these cover crops could vary widely depending on the climate, cropping system and the plant legume itself.

Non-Legume cover crops such as cereal grains, rye, and wheat have been used successfully in many different climates and cropping systems. There is also a growing interest in the usage of brassica and mustard cover crops due to their "bio-fumigation" characteristics. These plants release bio-toxic chemicals as they decompose. These chemicals have been found useful to reduce diseases, weeds and nematode pressure in the subsequent crops.

Non-Legume cover crops are mostly useful for:

- Scavenging nutrients (especially Nitrogen, N) that were leftover from previous crops
- Reducing and / or preventing erosion
- Producing a large amount of residue and adding organic matter to the soil
- Suppressing weeds

The common practice in Cameron Highlands is the use of non-legume cover crops like vetiver, chayote, chew, Kim, ryegrass, barley, brassicas, mustards, buckwheat, oats, rye, (e.g. Figure 2.12).

Figure 2.12: Napier Grass for Erosion Control

b) Mulching

Straw Mulch: In addition, to cover crops, another effective erosion control practices that are implemented almost anywhere is the use of straw mulch. This is to protect bare soil from raindrop impact by spreading of straw on the soil surface, but this type of mulching was not being practised in Cameron Highlands because of unavailability of the straw nearby.

Synthetic Mulch: This involved the use of synthetic materials to cover soil surface against erosion. Plastic materials are most commonly used for this purpose, and it is widely used in Cameron Highlands with a positive outcome (Figure 2.13).

Figure 2.13: Synthetic Mulching at Cameron Highlands

E. Tillage Practices

Tillage is the mechanical manipulation of soil to provide a suitable environment for crop growth, maintenance of infiltration capacity, aeration and weed control. It has an influence on soil structure, which in turn influences infiltration, aeration and soil erosion. Traditional or conventional tillage consists of ploughing and harrowing so that the soil is loosened, clods are broken and thus, burying crop residues.

Several modified tillage operations are available now, with the objective of providing improved soil-water-plant relations, reduce soil erosion and runoff, helping in moisture conservation and reducing the time and cost of tillage operation. The term minimum tillage describes the preparation of the seed-bed with minimum soil disturbance. The practice consists of opening the land to plant the seed and use chemical herbicides to control the weeds. One of the major benefits of minimum tillage is the increased residue left on the soil surface. Such residue is extremely effective in reducing erosion.

The tillage practices for BMPs erosion control are:

a) Strip or Zone Tillage

Strip or zone tillage consists of seedbed preparation by cultivating the soil in a narrow strip adjacent to the proposed planting row (Figure 2.14). The area in-between the rows is either left untilled or tilled in a different manner. It can be prepared in such a way that it helps to increase infiltration, surface detention, and control erosion. An example of zone tillage is broad-bed and furrow system.

b) Mulch Tillage for BMPs

Any tillage operation that will leave a substantial part of the residual vegetative material, such as leaves, stalks, straw on or near the surface as a protective cover is known as mulch tillage. The practice is also referred to as stubble mulching. The major benefits are:

- Reduce the beating action of raindrops and reduce splash erosion.
- Reduce inter-rill erosion by retarding surface flow.
- help in infiltration through maintenance of good soil structure, and
- Controlling soil temperature.

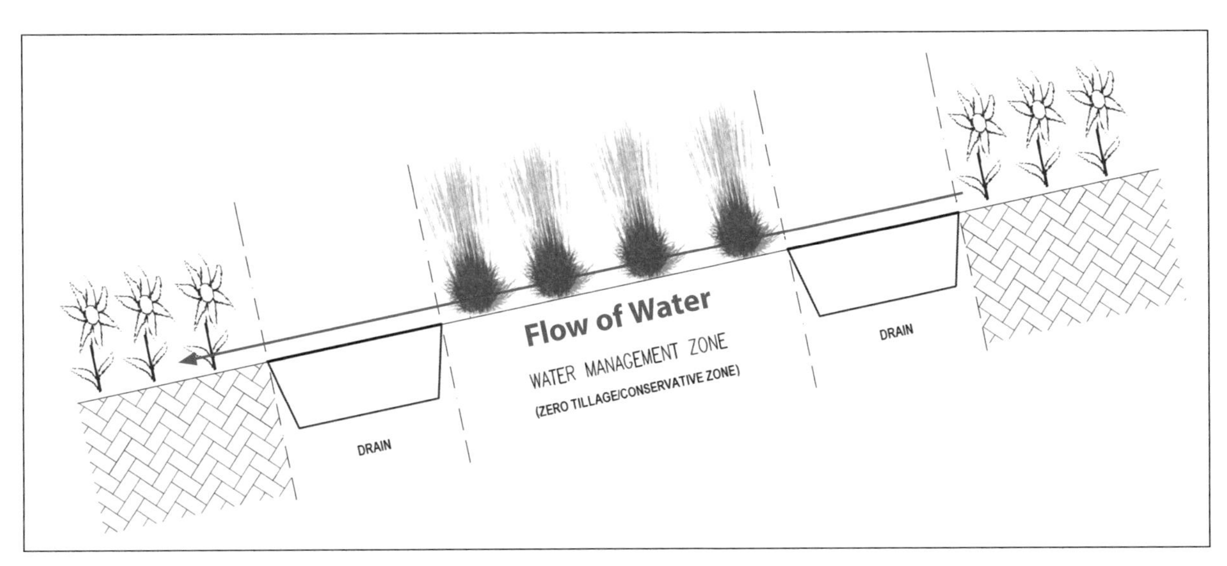

Figure 2.14: Zone tillage systems.

F. Soil Management Measures for BMPs

Soil management measures are concerned with ways of preparing the soil to promote dense vegetative growth and improve the soil structure so that it can offer more resistant to erosion. It helps not only to control soil erosion but helps to harness rainwater and conserve soil moisture. These measures aimed at increasing the resistance of the soil to erosion which includes; use of fertilizers and manures, subsoiling, drainage, conservation tillage, contour tillage, ridging and Tied ridge, minimum tillage, and no-till.

i. Tied Ridging

Tied ridging consists of covering the land surface with closely spaced ridges in two directions at right angles so that a series of rectangular basins are formed for BMPs. The purpose of such basins is to retain the rainwater until it infiltrates into the soil (Figure 2.15).

This system can be successful when carefully designed and constructed. The system can be successful on the level ground, or when the amount of water that can be stored in the basins and the amount infiltrating during the storm, is more than the worst storm likely to occur. Failure of a ridge, particularly on sloping land, can cause a series of failure of other ridges.

Tied ridging are more successful on permeable soils rather than on shallow soils. To prevent failure, they are constructed on grade with ties lower than the ridges so that the failure and runoff will be along each ridge and not down the slope. It is also advisable to back up the system with other measures like terraces or bunds.

Figure 2.15: Tied ridging
(Source: Briggs Irrigation Manual, 1991)

ii. Sub-Soiling

The practice consists of deep ploughing or chiselling using special equipment known as a chisel plough or subsoiler. The main objectives in sub-soiling are to break and shatter plough soles or another impermeable layer in the soil profile, loosen soil profile to considerable depths to permit deep leaching of accumulated salt in the upper layers, to deepen the effective plough zone depths for crop growth and to increase the infiltration rate and reduce run-off water. The effect of sub-soiling does not last long, and the operation has to be repeated after some years.

2.9.2 Engineering Measures

Engineering or physical measures depend on the manipulation of the surface topography to control the flow of water. Major engineering measures are:

A. Terracing,
B. Bunds
C. Drains for BMPs
D. Water Controlling Structures.

A. Terracing

Terracing is a method of erosion control accomplished by constructing a broad channel across the slope of rolling land (Figure 2.16). It entails breaking the slope length into a step-like pattern so that the slope is no longer continuous, thus reducing the slope length of overland flow and intercepting the runoff and conducting it to a safer outlet at a non-erosive velocity. In humid areas, terraces help to decrease the length of hillside slope thereby reducing rill erosion, prevent the formation of gullies and allow sediment to settle from runoff water, thereby improving the quality of surface water leaving the field. It can also be for recharging of shallow aquifers and as an aid in surface irrigation, particularly in the rice-growing area.

The main objectives of terracing are to modify soil slope, influence surface runoff and allow for the agricultural use of the steep slope. The purposes of terracing include Irrigation (use level terraces), soil management by reducing the slope of the area (use bench terraces), water management to absorb rains with emergency overflow (use contour bunds), reducing or controlling runoff (use graded channels terraces, ridging, tied ridging) and lastly, is for crop management (use intermittent terraces, orchard terraces, platforms, hillside ditches).

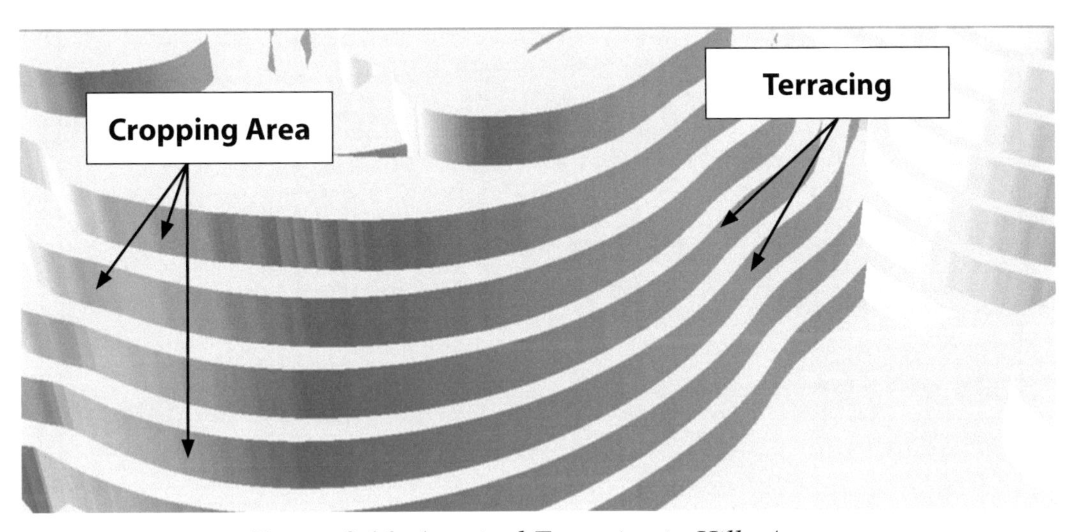

Figure 2.16: A typical Terracing in Hilly Areas

The ASAE (2011) classified terraces based on:

- Alignment
- Cross-section
- Grade
- Outlets.

Terraces can also be classified on the basis of purpose and use. Classification can be based on alignment;

- Non-parallel terraces
- Parallel terraces.

The non-parallel terraces follow the contour of the land. Some minor adjustments are frequently made to eliminate sharp turns and short runs. Parallel terraces meanwhile aid in the farming operation and should be installed wherever possible. Parallel terraces require more cut and fill volumes construction than non-parallel systems.

Bench terrace and Broad base terrace. There are three types:

- Sloping inward
- Sloping outward
- Level top

Bench terraces with slope inward are to be adopted in heavy rainfall areas where a major portion of the rain is to be drained as surface runoff. A suitable drain at the inward end of each terrace is to be provided to drain the runoff (Figure 2.17 (a)-(b) and Figure 2.18 (a)- (e).

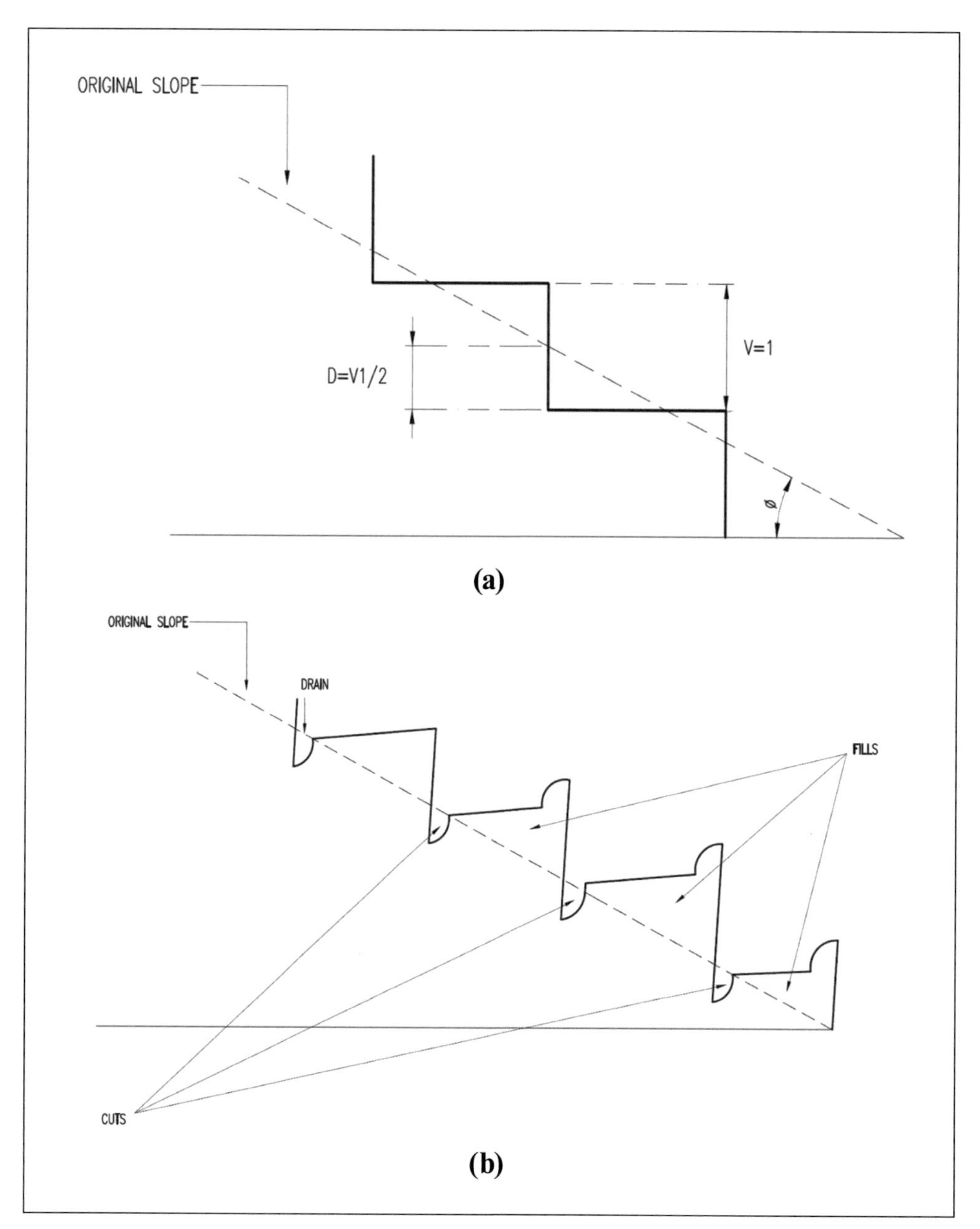

Fig 2.17: Bench Terrace: (a) Vertical and Horizontal Interval of Terracing; (b) Sloping Inward

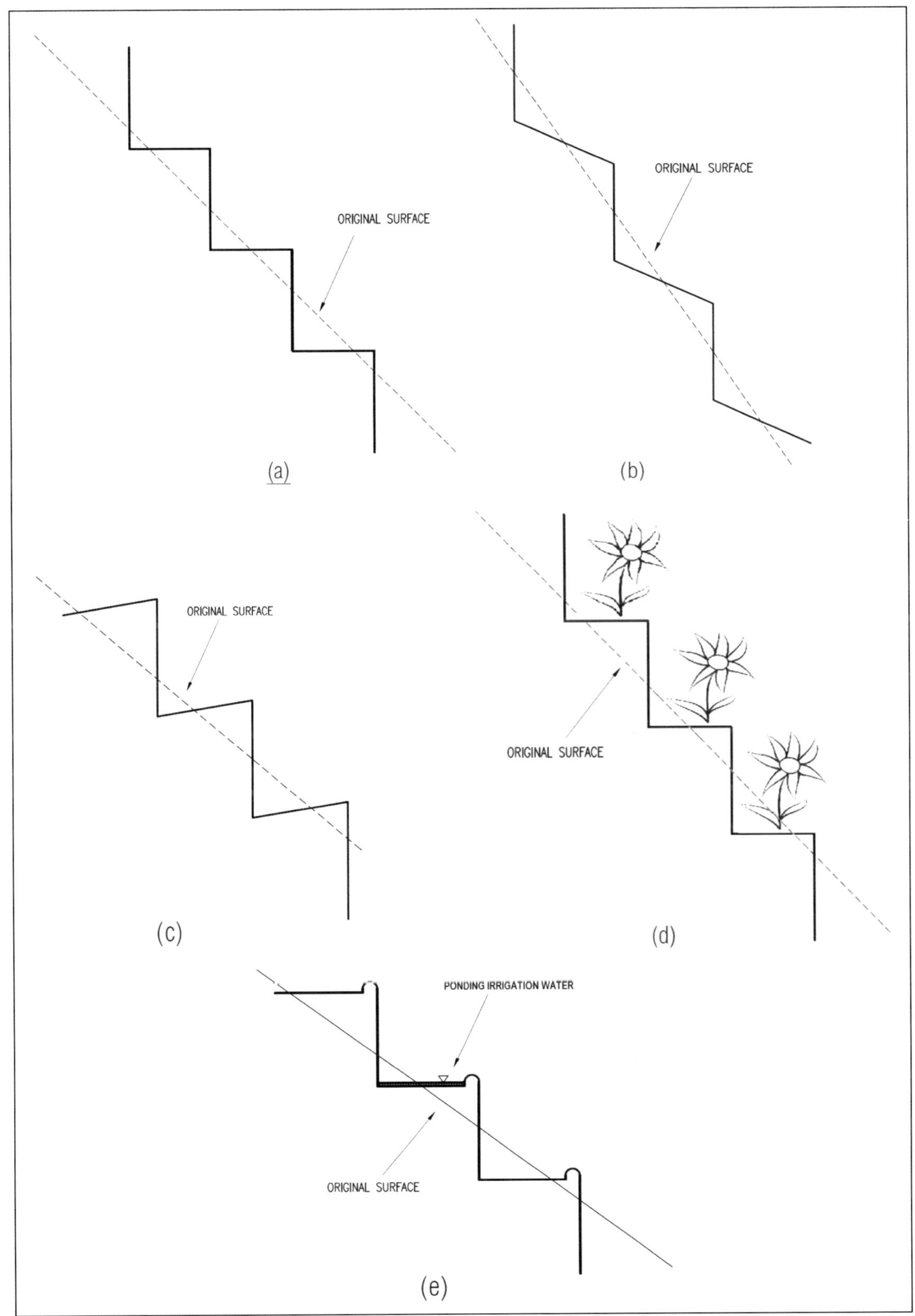

Fig. 2.18: Type of Terrace Design: a) Level bench; b) Outward sloping; c) Inward sloping; d) step terrace e) Irrigation terrace

Description of Terracing Design for BMPs

- Specifying the proper spacing and location of terraces
- Channel design with an adequate capacity
- Development of a farmable cross-section

The following considerations are important making the terrace design:

i. Soil characteristics such as soil depth, type, the average slope of the area
ii. Crop and soil management practices
iii. Climatic conditions
iv. The purpose for which the terraces are to be constructed

i. Determination of Terrace Spacing

Terrace spacing is generally expressed as the vertical interval (VI) between two successive terraces. The vertical and horizontal spacing of terracing can be determined from Equations 2.8 to 2.9 below;

$$L = \frac{V^{5/2} n^{3/2}}{(R-i)\sin^{3/4}\theta\cos\theta} \quad (2.8)$$

$$VI = L\sin\theta \quad (2.9)$$

Where,

L is slope length;

V is maximum permissible velocity,

n = Manning's constant, a value of 0.01 is recommended for bare soil;

R is rainfall intensity;

i is infiltration capacity,

(R-i) therefore is the peak rainfall excess;

θ is the slope angle.

The Manning's constant;

$$n = \frac{r^{3/2} s^{1/2}}{v} \quad (2.10)$$

Where;

r = hydraulic radius, ratio of area to wetted perimeter (m)

s = slope the channel

v = velocity (m/s)

n = roughness coefficient

VI can also be expressed in terms of terrace width (W) and field slope, in per cent

$$VI = WS/100 \qquad (2.11)$$

Another Empirical relation is;

$$VI = aS + b \qquad (2.12)$$

Where

VI is vertical height;

a is constant which depend on geographical factors;

b is a constant of soil-erodibility and cover condition during the critical erosion period;

S is per cent average land slope above the terrace.

The values of a and b vary for one geographical location to another, and the form of expression of the relationship varies.

ii. Terrace Grades for BMPs

The gradient in the channel must be sufficient to provide adequate drainage while removing runoff at non-erosive velocities. The minimum slope is desirable from the standpoint of soil loss. The grade may be uniform or variable. In the uniform-graded terrace, the slope remains constant throughout its entire length. A range of grade of 0.1 to 0.6 % is recommended for BMPs depending on the climatic and soil factor, and an average of 0.4% is common in many regions of the world. Generally, steeper grades are recommended for impervious soils and short terraces.

The variable-graded terrace is more effective because the capacity increases towards the outlet with a corresponding increase in runoff. The grade may vary from a minimum at the upper portion to a maximum at the outlet end, to reduce the velocity in the upper reaches. This reduction in velocity provides for greater absorption of runoff and effective deposition of sediment. Variable gradient makes flexibility in design possible.

iii. Terrace Length

Factors that influence terrace length include:

- Size and shape of the field
- Outlet possibilities
- The rate of runoff as affected by rainfall and soil infiltration
- Channel capacity

The number of outlets should be minimum and consistent with good layout and design. Incredibly long graded terraces should be avoided for BMPs; however, the extended length may be reduced in some terraces by dividing the flow midway in the terrace length and drain the runoff to outlets at both ends of the terrace. The length should be made such that erosive velocities and large cross-sections are not encouraged. The maximum length of graded terrace generally ranges between 300 to 500 m, depending on the local conditions

iv. Terrace Construction for BMPs

Table 2.11: Design criteria for the construction of a terrace

Parameters	Description / Criteria
Equipment	• Terracing machines include a bulldozer, pan or elevating scraper, motor patrol or blade grader and elevating grader can be selected depending on the size of work. Smaller equipment, such as moldboard and disks ploughs are suitable for slopes less than 8%.
Vertical / Horizontal Intervals	• The vertical and horizontal interval should be calculated from the above mathematical formula
Cut and Fill	• Cut and fill operation is continue down the slope length

v. Limitation

Terraces or bench terraces should not be recommended on the following conditions for BMPs:

- A sandy or rocky soils, non-cohesive or highly erodible soils, or decomposing rock including other depositional materials.
- On recently soil-cuts and soil-filled up segment – this is because the cut part could be infertile and the fill may not have settled properly.
- On soft-rock laminations in thin layers oriented so that the strike is approximately parallel to the slope face and the dip approximates the staked slope line.
- Benches terraces may cause sloughing if too much water infiltrates in the soil and are effective only where suitable runoff outlets are available
- Avoid benching, if possible, in areas where there is potential for rock-fall slide problems.

vi. Basic Design Requirement For Bench Terrace

Before going directly to the design of bench terraces designers should understand the following design basics requirements;

a. Design terraces according to the needs of farmers, crops, climate, and tools to be used for farming.
b. Use simple arithmetic and a step-by-step approach to design, this to mean that start on simple parameters that you easily determine. For example, using the land slope and the width of the bench (flat part) as two starting points, the design proceeds step by step with basic arithmetic that can be easily understood by field workers, land users, or farmers.
c. Design of bench terraces should be in such a way that the volumes of cut and fill are to be equal for minimizing construction cost.

A. Bunds for BMPs

Bunds are small embankments constructed across the slope of the land. They are used in high rainfall or low rainfall conditions, varying soil types and depths for soil and water conservation and afforestation. The bunds may be made with a trench which helps to break the slope length, reduce surface runoff velocity and consequently retard its scouring action and carrying capacity. The water retained in the trench helps in conserving the moisture and provide advantageous sites for sowing and planting (Figure 2.19).

Description;

- The bunds are considered contour bunds when they are constructed on the contour and graded bunds when a grade is provided to them.
- Contour bunds may be continuous or interrupted. The interrupted bunds may be staggered or inline.
- Continuous bunds are essentially used for moisture conservation in low rainfall areas and required a careful layout. Interrupted bunds are adopted in high rainfall areas.
- Even though bunds may be used on all slopes, bunds on slopes exceeding 20% are not advisable either technically or economically and are usually not cultivated.
- This is what makes them different from broad-based terraces which are cultivated.
- Bunds are suitable for lands slopes from 2% to 10%. Soil erosion in an area with less than 2% may be controlled using biological measures while bench terraces should be used for slopes beyond 10%.
- If bunds are to be used, they must be at very close range, which will limit the crop cultivation area in such land (Figure 2.19).

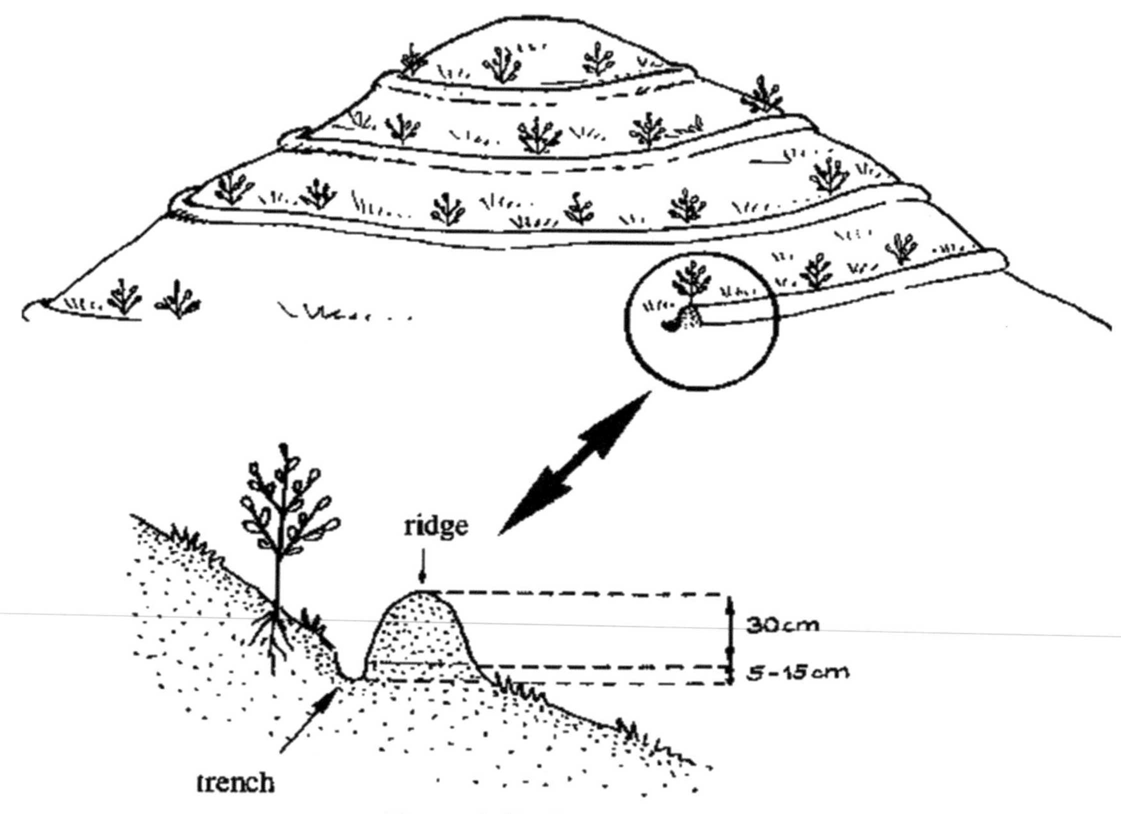

Figure 2.19: Contour Trench
(Source: ILO - UNDP, 1993)

Contour Trench BMPs

- Firstly, find the contour lines and mark them with pegs.
- Remove thc topsoil and save it above the terrace. Mixing fertile topsoil with infertile subsoil should be avoided.
- Dig the step. Place the excavated material downhill to extend the terrace.
- Replace the fertile topsoil.

i. Design Specification of Bund

A band consists of following parameters to be determined under design:

- Choice of bund
- Spacing
- Size
- Side slopes
- Alignment of bund

Choice of Bund: The choice of contour or graded bunds depends on the rainfall, soil and the type of outlets. In large and medium rainfall areas, the grade of the bunds maybe 0.2 to 0.3% towards the outlets.

The spacing of the bunds: The following principles should be considered in spacing bunds:

The seepage zone below the upper bund should meet the saturated zone of the lower bund. The bunds should check the water at a point when the water attains erosive velocity as well as not inconvenience agricultural operation.

To determine the bund spacing, equation 2.5 is can be used depending on the nature of rainfall in the area. Equations 2.6 and 2.7 are applicable at medium and heavy rainfall zone and in low rainfall zone respectively.

$$V.I = \frac{S}{a} + b \qquad (2.13)$$

Where;

V.I. = vertical interval between co nsecutive bunds (cm)

S = land slope in percent

a and b are constants depending on the soil and rainfall characteristic of the area.:

$$V.I = \frac{30S}{3} + 60 \qquad (2.14)$$

$$V.I = \frac{30S}{2} + 60 \qquad (2.15)$$

ii. Construction of Bund for BMPs

- Construction of bund should start with bund nearest to the ridge and continue down the valley (Fig 2.20). This will ensure bunds protection if rains occur during construction.
- The burrow pits for the soil should generally be located at the upstream side of the bund. It should be of a uniform depth of 30 cm for BMPs and width may vary as necessary.
- The burrow pits are to be continuous and no breaks should be left, and should not be located in a gully or depression, and when the soil is dug, all the clods should be broken in the burrow pit itself before putting it in the bund.
- The earth should be put in layers of 15 cm and consolidated by trampling. The bund section should be finally shaped, trimmed and slightly rammed to the side surfaces.

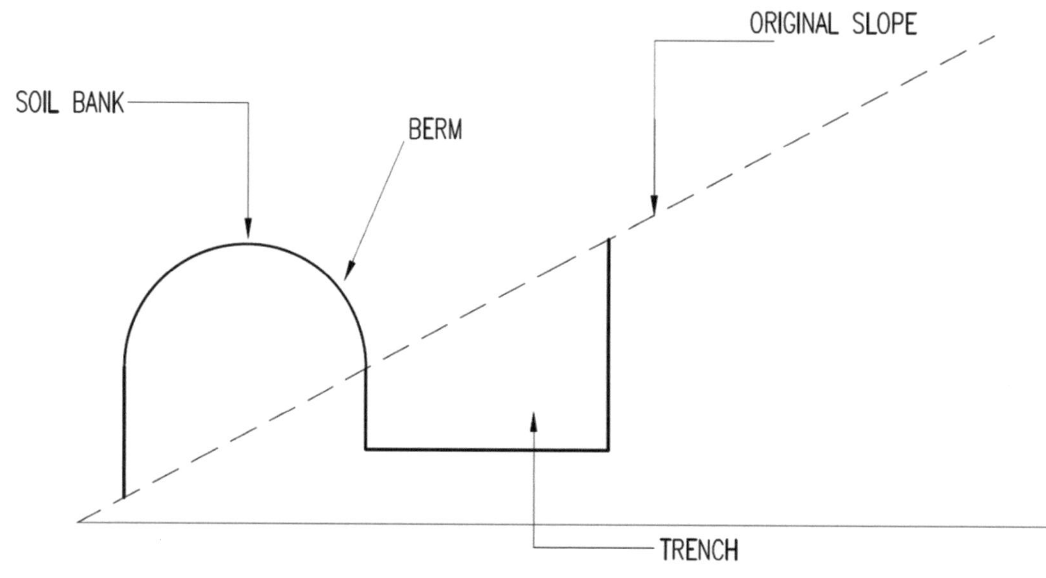

Fig 2.20: A typical Bund

iii. Alignment of Bunds for BMPs

- As far as possible, the bunds should be straight, sharp curves should be avoided for BMPs and should cross a gully or depression at right angles.
- Deviation of up to 15 cm in crossing a ridge and up to 30 cm in crossing a gully is allowed for BMPs, and when the bund crosses a depression, provision should be made to strengthen the bund section.
- In the case of graded bunds, extended length of bunds should be avoided. The length of bund in the direction of water should not exceed 350 m for BMPs
- Ramps for passage of equipment and cattle may be provided where necessary.

B. Drains for BMPs

Drains are open channels constructed along the slope of agricultural lands and act as an outlet for runoff from bunds and terraces. The purpose of a drain in a conservation system is to convey runoff at a non-erosive velocity to a suitable disposal point. A drain, therefore, must be carefully designed to meet the BMPs criterion. Its dimensions must have sufficient capacity to carry the peak runoff from a storm of about 10-year return period.

There are three types of drains as follows;

a) Diversion channels
b) Terrace channels
c) Grass-covered drain

Description

- Diversions channels are placed upslope of areas where protection is required to intercept water from the hillside.

- They are built across the slope and given a slight grade, to convey the intercepted runoff to a suitable outlet.
- Terrace channels are placed upslope of terrace bank to collect runoff from the inter-terraced area.
- They are also built across the slope with a slight grade to suitably convey the runoff to the outlet. The grass-covered drain is used as an outlet for diversion and terrace channels, they run downslope at grades of the sloping area and empty the runoff water into river systems or other outlets.
- They are usually located in natural depressions on the hillside where possible. Occasionally, natural channels are reshaped to serve as a grass-covered drain, may also serve as emergency spillways for farm ponds.

Grass-covered drains are recommended for BMPs at slopes up to 11 degrees. On hillsides with alternating gentle and steep sections, a grass waterway with drop structure on the steeper slope should be used. The selection of grasses used should take into account soil and climatic condition that can establish a dense cover very rapidly. On the steeper slope of up to 15 degrees, the channel to be used should be a stone peach or concrete-lined.

i. Establishment of Grass-Covered Drain:

The design of the grass-covered drain is similar to the design of an irrigation channel (the design procedures are based on the principle of open channel hydraulics). However, it is more complicated than the design of channels lined with concrete or other stable materials because of the variation in the roughness coefficient with the depth of flow, stage of vegetal growth, hydraulic radius and velocity. The basic aim is to control runoff, which can result to channel or gully formation. So the focus is:

a. Reduction of peak flow rate by full utilization of field protection practices
b. Provision of a stable channel that can handle the discharge flow

Description

- Stabilization of the drain can be best accomplished by providing vegetal protection for the channel to limit the flow velocities that the vegetation can stand.
- For large runoff volumes or steep channels, it may be necessary to supplement the vegetated watercourse with permanent gully control structures.
- Vegetated drainage should not be used for continuous flows such as discharges from tile drains, as for prolonging wetness in the waterway results in poor vegetal protection.
- In the design of a vegetated watercourse, the functional requirement should be meet for sustainable operation.
- The capacity of the drain should be based on the estimated runoff from the contributing drainage area.
- The 10-year return period is a sound basis for BMPs vegetated-covered drain design, except in flood spillways for dams where long return periods is required.

For a more comprehensive design criteria, please refer to MSMA 2nd Edition, 2012.

C. Erosion Control BMPs for Sheltered Farm

Sheltered farming activities are quite different and have more control over unsheltered farming described earlier. In this case, rainfall is not in contact with both soil and crop and intercepted by roofing (Figure: 2.21). The more intercepting of rainwater, the larger the volume of runoff is collected. Hence, for BMPs, this runoff must be managed and transported out of the field safely and at a nonerosive velocity.

Description for BMPs

a. Entire cropping area should be covered with a roof tilting towards a particular drainage outlet.
b. A number of drainage outlet may be required depending on the size of drains and runoff volume.
c. The drain (usual pipes) should be guided down to the level ground. Then, horizontal pipes should be used to convey the water drainage system.
d. The lateral pipes (drains) should be provided to convey runoff water from farm efficiently.
e. Usually, a reservoir is provided to collect all the runoff water before discharging to a drainage system.
f. The drains size should be designed to contain the maximum amount of runoff economically.

i. *Design Of Drain Size For BMPs*

The design of the drainage system in the sheltered farm is very important in order to safely convey the runoff at a nonerosive velocity. Fig 2.20 shows a typical example of a sheltered farm at Cameron highlands hilly area where the runoff from the roofing is channel down the slope.

Fig 2.21: Drainage system of the sheltered farm in Cameron Highlands

For a roof with surface area 20 m x 50 m, will generate a discharge of 20 x 50 x n (m3/s), where, i is the rainfall intensity. Therefore;

Discharge $Q = W L I$ .... (2.16)

Where;

Q	=	discharge (m3/s)
W	=	width of sheltered (roofing) in m
L	=	length of sheltered (roofing) in m
i	=	rainfall intensity (mm/ha)

ii. Gutter Drain Collector, and Perimeter Drain

The purpose of a perimeter drain system is to disperse the water collected from the roof surface away from the foundation wall of a house. The roof drainage system and the perimeter drain system need to work together to manage the water around a shelter. The water flows from the gutters to the downspouts, then ideally the downspouts tie into a perimeter drain system. This will reduce the amount of water pushing against the foundation walls, also known as hydraulic pressure.

Downspout material is relatively inexpensive. There are some fancier products that roll out when it rains. Installation of a perimeter drain system is costly; a ditch needs to be dug around the shelter, back-filled with drain rock and drain pipe then topped and re-landscaped.

Problems observed with sheltered farms in Cameron Highlands;

- No lateral drains to convey the runoff to the main drain outside the farm. This resulted in flooding some areas of farms and washing away both soil and nutrients.
- Some of the drains were chocked up by waste generated from the farms.
- There were no economic considerations in some drainage designs for sheltered farms, as such the drain sizes are found to be too large.

2.10 RAINWATER HARVESTING FOR AGRICULTURAL AREAS

Rainwater harvesting is one major key to meeting the global millennium goal on water issues. The water collected can be concentrated in the soil profile or the artificial reservoirs such as pond, basin, or tank and used for various purposes, ranging from crop production to water supply for domestic, animals and other productive uses. RWH technique is, therefore, a collection and concentration of direct rainfall or surface runoff for productive purposes, instead of runoff being left to cause soil erosion, on-farm direct use of rainwater in rain-fed agriculture and irrigation through practices which enhance soil water use efficiency (Figure 2.22).

2.11 MAIN COMPONENTS OF RAINWATER HARVESTING (RWH) SYSTEM

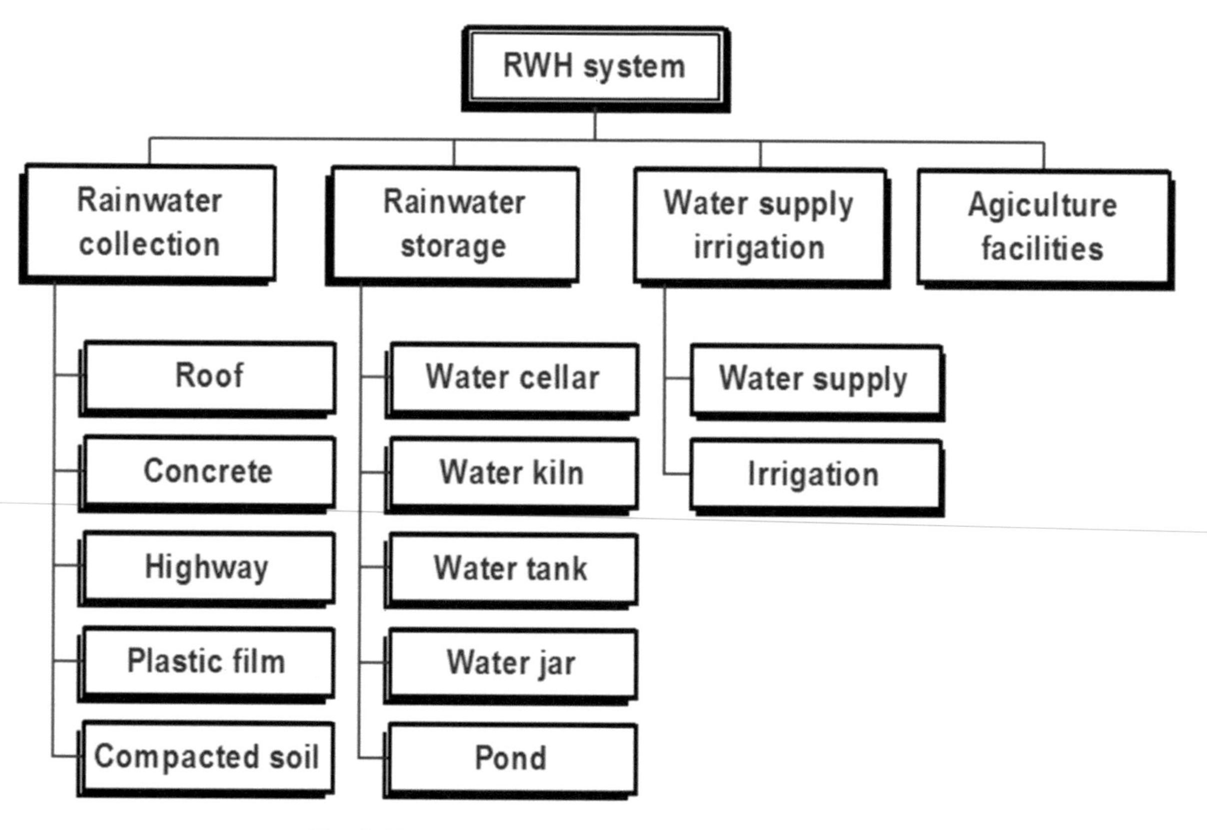

Fig 2.22: Micro catchment techniques of RWH

Rainwater-harvesting system can be classified into the following basic categories:

a) Rooftop rainwater harvesting technique: used to harvest water for domestic and animal water needs.
b) Micro-catchment techniques: In-situ water harvesting for crop production (Simple RWHS).
c) Macro-catchment techniques: can be used to harvest water for animal and agricultural use (Complex RWHS).

2.12 ROOFTOP RAINWATER HARVESTING TECHNIQUE

For more than three centuries, rooftop catchments and cistern storage have been the basis of domestic water supply in many places especially in rural areas where the rainwater harvesting is an important source of water supply for domestic purposes and irrigation purposes.

A rainwater harvesting system consists of three basic elements: a collection area, a conveyance system, and storage facilities. The collection area, in most cases, is the roof of a house or a building (Figure 2.23). The effective roof area and the material used in constructing the roof influence the efficiency of collection and the water quality. A conveyance system usually consists of gutters or pipes that deliver rainwater falling on the rooftop to cisterns or other storage vessels. Both drainpipes and roof surfaces should be constructed of chemically inert materials such as wood, plastic, aluminium, or fibreglass, to avoid adverse effects on water quality issues.

The water ultimately is stored in a storage tank or cistern, which should also be constructed of an inert material. Reinforced concrete, fibreglass, or stainless steel are suitable materials and in most cases use for this purposes. Storage tanks may be constructed as part of the building or maybe built as a separate unit located some distance away from the building.

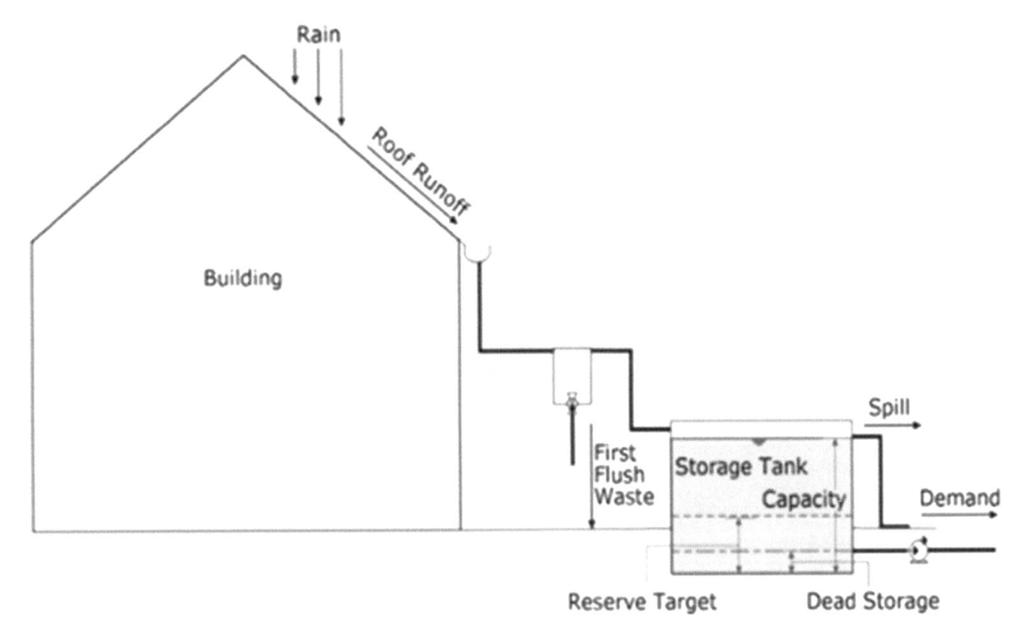

Figure 2.23 RWH from top Roofing

Seven primary components of a rainwater harvesting system are;

- Contributing Drainage Area (CDA)
- Collection and conveyance system (e.g. gutter and downspouts)
- Pre-screening and first flush diverter (Pretreatment)
- Storage tank
- Water quality treatment
- Distribution system
- Overflow, filter path or secondary stormwater retention practice

2.13 RAINWATER HARVESTING FOR AGRICULTURE FARMING - BMPS

Basic Requirements;

1. CDA surfaces should be smooth and composed of non-porous materials so that the drainage will be more efficient.
2. The material of CDA should be carefully selected so that it does not leach toxic chemicals. Certain types of rooftops and CDAs, such as asphalt seal coats, tar, and gravel, painted roofs, galvanized

metal roofs, sheet metal, or any material that may contain asbestos may leach trace metals and other toxic compounds and should be avoided.

3. Pre-treatment is required to keep sediment, leaves, contaminants, and other debris from the system and hence prevent organic build-up.
   - Leaf screens separate leaves and other large debris from rooftop runoff.
   - First, flush diverters can be used to remove smaller contaminants such as dust, pollen, and bird and rodent feces.
   - Roof washers are placed just ahead of storage tanks and are used to filter small debris from harvested rainwater.
   - Vortex filters can provide filtering of rainwater from larger CDAs.
4. The size of the rainwater harvesting system should be determined through design calculations.
5. Materials used to construct storage tanks should be structurally sound and suitable for potable water or food-grade products.
6. Tanks also must be water-tight to prevent water leakage and damage the building (i.e. house) foundation.
7. Ensure that the available space is adequate and suitable to house the storage tank and any overflow.
8. Site topography and storage tank location should be also be considered as it will affect pumping requirements.
9. The hydraulic head should be determined either using gravity or pump.
10. Storage tanks can also be located under the ground under the following condition;
    a. The tank should be buried above the water table, so it does not praise the water table and hence can avoid the flooding.
    b. All nearby underground utilities should be avoided during the installation of the underground tanks.
    c. Underground storage tanks should be placed in areas without vehicle traffic or heavy loading
11. The storage tanks can be very heavy, and hence it should be placed on a gravel or sand pad.
12. The collection and conveyance system consists of the gutters, downspouts, and pipes that channel rainfall into storage tanks. Aluminium, round- bottom gutters and round downspouts are generally recommended for rainwater harvesting.
13. If a pump is needed to distribute the water, type of pump (i.e. an external, submersible pump) and its size should be determined properly. Some pump designs may require a backup water supply to ensure proper operation of the pump during low water level periods.
14. Limestone or other materials should be added in the tank if the rainwater is acidic.
15. Overflow mechanism should be installed if there is a need to spill away from the excess water, especially during a storm event. The overflow pipe should be screened to avoid animals coming into tanks. The spillway should be directed to an acceptable flow path that will not cause erosion.
16. Storage tank overflow devices should be able to direct overflow away from buildings foundation approximately 10 feet to avoid soil saturation.

i. Micro Catchment Techniques

In micro-catchment techniques, runoff flows directly into the cropping area. The system collects rainwater and allows it to seep in the soil, where plant roots can reach it. The techniques are usually staggered in alternate rows so that overflow from one row runs into the next down the slope (Figure 2.24). In Micro catchment techniques the catchment is not more than three times the size of the cropping area. Its size depends on the amount of water that needs to be retained.

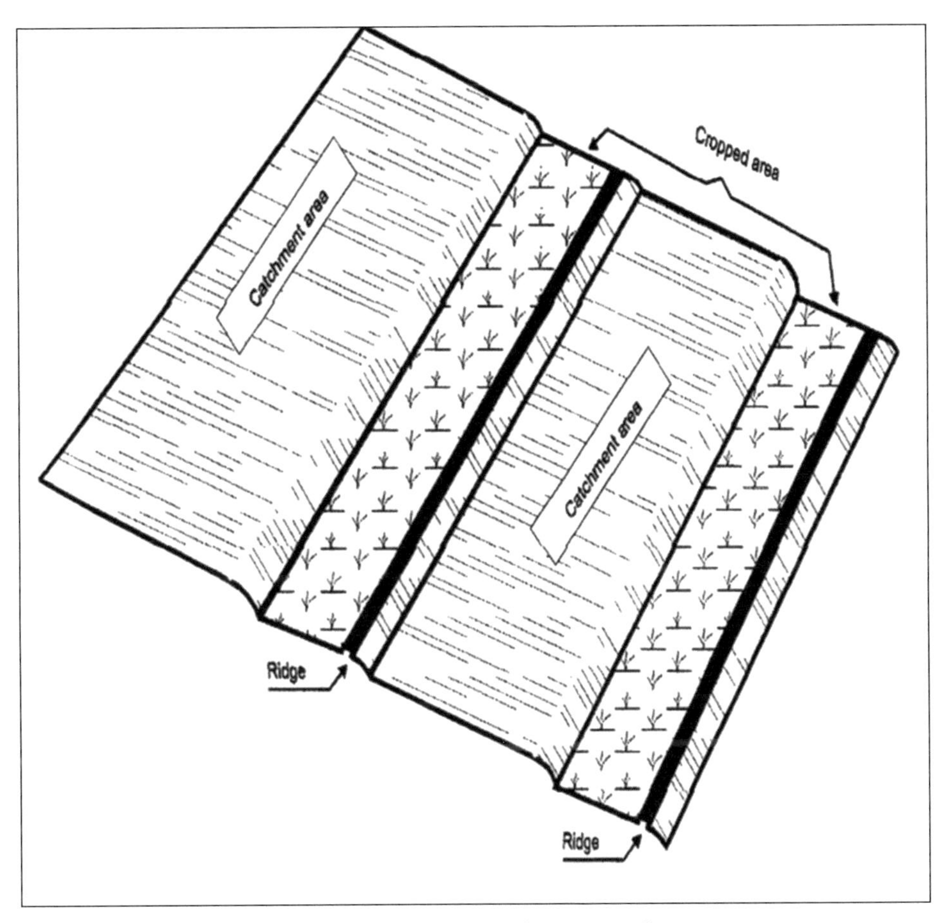

Figure 2.24: Micro catchment techniques

ii. Macro-Catchments (In-Situ) Rainwater Harvesting

This technique is a large-scale water harvesting system to collect runoff water over a distance of 150 m (external catchment system) for BMPs. It entails harvesting direct rainwater and runoff on agricultural fields. Channel usually connects the catchments with the cropping area. It covers a much larger area than micro-catchments. The water is collected in catchments, and then it is transported to a separate place where it is stored and used (Figures 2.25 and 2.26). The catchment area is usually several times larger than the cropping area and maybe hundreds of meters away from the cropping area.

The rainwater catchment area is a natural or man-made unit draining runoff water to a common point. It could be; Roof catchment, Rock catchment, paved ground catchment, Roads and Sand dams/dry river bed, etc.

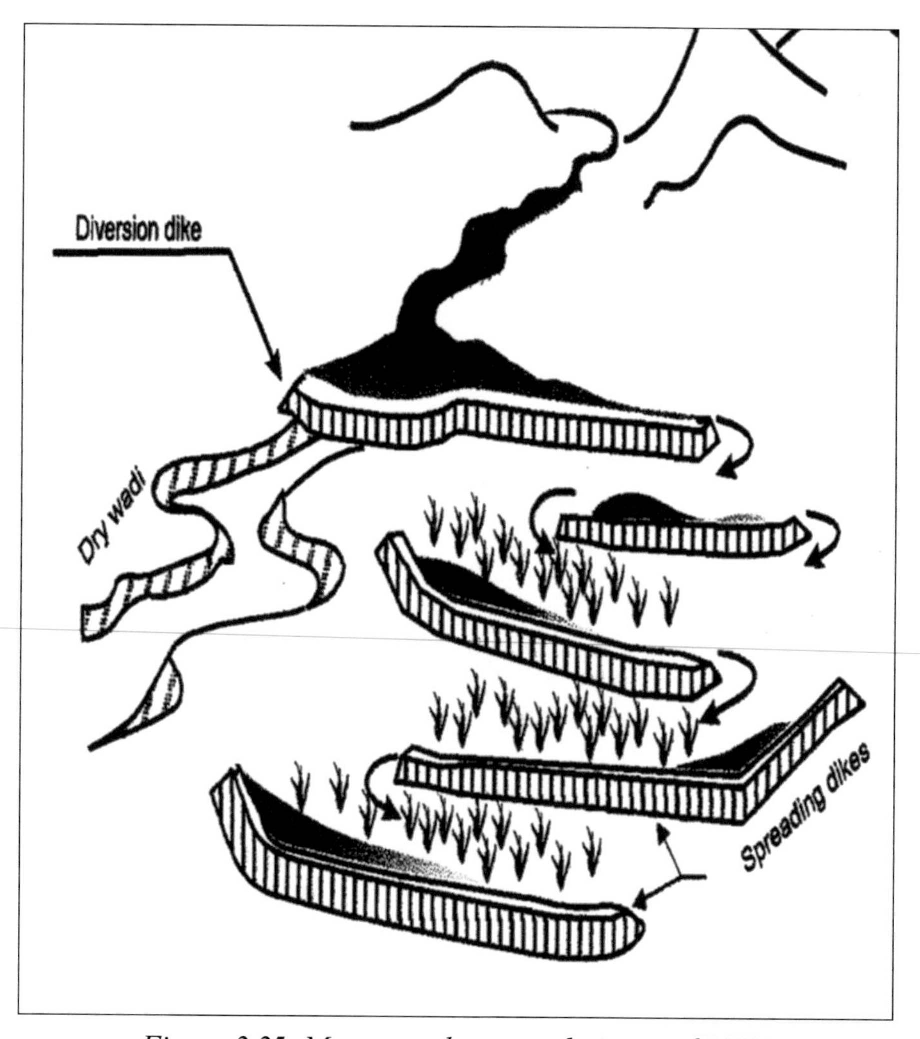

Figure 2.25: Macro catchment techniques of RWH

Fig 2.26 Example of Macro Catchment techniques for Cress water cultivation

Factors of Considerations for BMPs;

- The catchment area must be able to collect sufficient runoff water to meet the designed storage capacity or to meet the users' needs.

- The runoff water collected from the catchment area should be easily diverted to the tank.
- The catchment area should be located sufficiently away from pollution sources (toilets, animal sheds), and must be protected from contamination.
- The catchment must generate as little sediment as possible. Use suitable soil conservation measures to reduce the amount of silt that is carried into the tank.
- Selecting a suitable site for the water tank construction is essential, as it will be permanent.
- Do a preliminary survey before designing and constructing the tank. If possible, compare two or more locations before selecting the most practical site.
- Compare two or more locations before selecting the most practical site.

Benefits of Macro rain-water harvesting

i. The harvested water is stored in the soil, thus increasing the soil moisture content in the plant root zone.
ii. In-situ RWH help to reduce soil erosion since the runoff on the fields are controlled and harvested
iii. It is one way of managing the water resource of the small watershed as it helps to increase groundwater recharge and prevent waterlogging and flooding.
iv. In-situ RWH involves reforming the land to retain as much rainwater in the soil as possible.

Various types of micro-catchment rainwater harvesting techniques which are already adopted by many farmers in multiple countries and/or which have good results include:

a. Terracing
b. Negarim Micro catchment
c. Semi-circular bunds
d. Contour ridges
e. Contour stone bunds
f. Contour soil bunds
g. Broad bed and furrow system
h. Ridging and tied ridging
i. Zay pits
j. Runoff harvesting from the road and Root storage basins

2.14 SEDIMENT CONTROL BMPS

2.15 SILT FENCE

A silt fence is a temporary sediment barrier consisting of filter fabric stretched across and attached to supporting posts, entrenched, and, depending upon the strength of fabric used, supported with plastic or wire mesh fence (Fig 2.27). The main function of silt fences is to trap sediment by intercepting and detaining small amounts of sediment-laden runoff from disturbed areas.

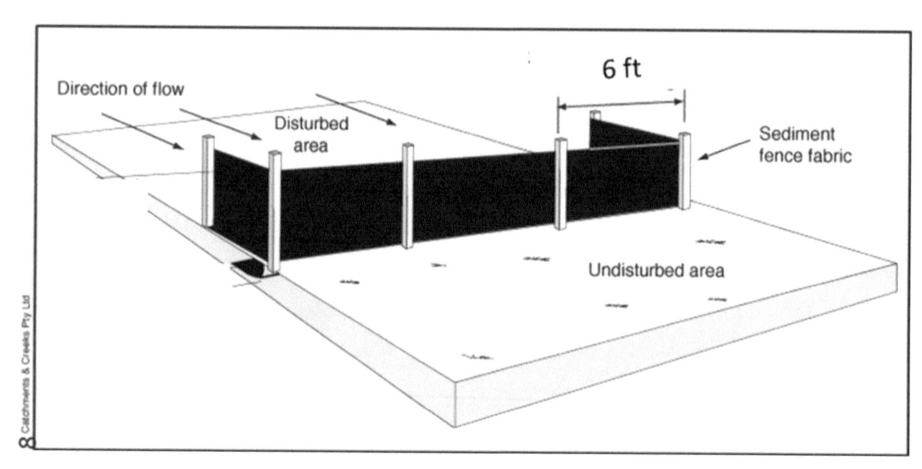

Fig 2.27: A typical Silt Fence

A. Materials for Silt Fence Construction

1. Filter fabric should be specified so that it can retain the soil, yet it has openings large enough to permit drainage and prevent clogging. The equivalent opening size should be selected based on the following criteria:
   - If 50 per cent or less of the soil, by weight, will pass the U.S. Standard Sieve No. 200, select the equivalent opening size (EOS) to retain 85 % of the soil.
   - For all other soil types, the EOS should be no larger than the openings in the U.S. Standard Sieve No. 70.
2. Silt fence fabric should be woven polypropylene with a minimum width of 36 in. and a minimum tensile strength of 100 lb force.
3. If the fabric selected does not have sufficient strength, the fence should be supported by a plastic or wire mesh.
4. Wood stakes should be commercial quality lumber. Each stake should be free from decay, splits, cracks or other defects that would weaken the stakes.
5. Staples can be used to fasten the fence fabric to the stakes.

B. Location

1. The silt fence should be located at areas where sheet flow occurs.
2. Should not be located in streams, channels, or anywhere flow is concentrated.
3. Silt Fence should not be located below slopes, as it may induce landslides.
4. The silt fence should be installed along a level (flat) contour to avoid water ponding along the silt fence.
5. Ensure that there is sufficient room behind the fence to make way sediment removal equipment.
6. The ends of the filter fence should be turned uphill to prevent stormwater from flowing around the fence.
7. The maximum slope perpendicular to the fence line should be 1:1. For steeper slopes, additional protection such as a chain-link fence or cable fence should be installed

C. *Installation*

Table 2.12: Criteria for the construction of Silt Fence

Items	Description / Criteria
Installation criteria	• A trench should be excavated approximately 6 in. Wide and 6 in. Deep along the line the proposed silt fence. • Bottom of the silt fence should be planted in the soil about 12 in. • Posts should be spaced a maximum of 6 ft. Apart and driven securely into the ground a minimum of 18 in. Or 12 in. Below the bottom of the trench. • In case mesh is used to support the fence, then the mesh should extend into the trench. • The trench should be backfilled with compacted native soil material. • The silt fence should be constructed at least 3 ft. from the toe of a slope so that it will be more effective and easy to maintain.

For more comprehensive design criteria, please refer to Chapter 12 of MSMA 2nd Edition, 2012.

D. *Inspection and maintenance*

1. The silt fence should be inspected weekly during the rainy season and at two-week intervals during the non-rainy season.
2. Repair or replace if there are any undercut silt fences, split, torn, slumping, or weathered fabric.
3. Sediment that accumulates in the silt fence must be periodically removed to maintain its effectiveness.

2.16 CHECK DAMS

Check dams are temporary or permanent linear structures placed perpendicular to concentrated flows such as in drainage ditches, channels, and swales to reduce flow velocities and prevent channel down-cutting. Some sediment trapping may occur during low flows. Check dam materials may include rock, fiber logs (e.g., wattles), triangular sediment dikes, sandbags, and other materials or prefabricated systems. Straw/hay bales and silt fences should not be used to check dam applications, as they are not intended for concentrated flow areas. Check dams are used to regulate flow velocities, reduce scour erosion, and trap small quantities of sediment along higher-risk ditches and channels that have slopes higher than 10 per cent and soil types conducive to erosion (e.g., sandy/silty soils). They are appropriate for both temporary and permanent ditches and swales. While most flatter and shorter channels (i.e., slope less than 3 per cent, length less than 200 feet) generally do not need check dams if they are stabilized immediately after construction (i.e., with sod, or seed and the appropriate rolled erosion control product), longer and steeper ditches can benefit from check dam installations. When evaluating the use of check dams for a particular site, consider the following.

Table 2.13: Criteria for Check Dams

Items	Description / Criteria
Installation	• Rock from a temporary check dam can be spread into a ditch and used as a channel lining when the check dam is no longer necessary. • Removal may be costly for some types of check dams.
Drainage area	• Check dams are suitable only for a limited drainage area (generally 10 acres or less). • Check dams are intended for use in small open channels, not streams or rivers.
Effects	• The hydraulic capacity of the channel can be reduced when check dams are in place. • Check dams may create turbulence downstream, causing erosion of the channel banks. • Ponded water may kill grass in grass-lined channels. • Check dams may be an obstruction to construction equipment.

For more comprehensive design criteria, please refer to Chapter 12 of MSMA 2nd Edition, 2012.

2.17 SABO DAM

Sabo dams sometimes called debris dams or check dams, which can be used to drastically reduce the impact of debris or sediments on downstream channels and water reservoirs. The primary purpose of the Sabo structures is to reduce the excess sediment discharge to prevent river degradation further downstream and to enable in the downstream flow in full capacity. More recently the Sabo works are being used to the control of debris flows in steep unstable topography with frequent volcanic activities and earthquakes.

Generally, the term "Sabo works" refers to the mountain protection system against disasters such as sediment flow, debris flows or landslides. Early Sabo works were undertaken during the 17th and 18th centuries. The most common type of Sabo dams is the vertical concrete wall (Fig 2.29).

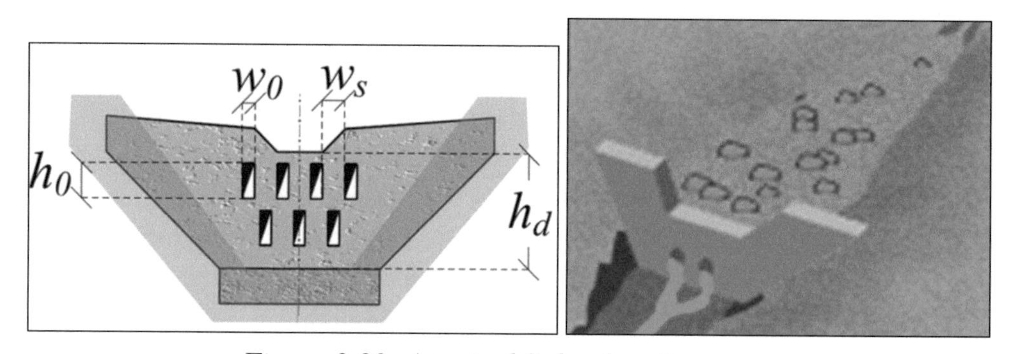

Figure 2.29: A typical Sabo dam Structure

Description Sabo for BMPs

- The downstream face of the dam is nearly vertical, followed by a short stilling structure.
- In steep topography, the downstream channel should be stepped to contribute to further energy dissipation, in a fashion somehow similar to stepped spillways.
- The Sabo dam height is to be kept in the range from 3 to 15 m typically.
- The structure has the initial purpose to trap sediment material and to reduce the slope of the upstream catchment when the reservoir is filled.

Other similar dams that serve the purpose as that of Sabo check dams include permeable check dams, tubular grid dams, slit dams, and overflow stepped weirs. Permeable check dams are designed to trap small to medium size debris. They do not hold water. In forest areas, the permeable dam may be made of steel grids.

2.18 SEDIMENT TRAP

A sediment trap is a temporary containment area that allows sediment in a collected storm or runoff water to settle out during infiltration or before the runoff is discharged through a stabilized spillway or dewatering pipe. Sediment traps are commonly used at the outlets of stormwater diversion structures, channels, slope drains, construction site entrances, vehicle wash areas, or other runoff conveyances. For more comprehensive design criteria, however, please refer to Chapter 12 of MSMA 2nd Edition, 2012.

Table: 2.14: Criteria for sediment trap

Items	Description / Criteria
Location	• Make sure to situate sediment traps at suitable locations for easy access by maintenance personals.
Side Slopes	• While excavating an area for a sediment trap, make sure the side slopes are not steeper than 2:1 and the embankment height not more than 5 feet from the original ground surface.
Outlet	• Install dewatering pipe, if necessary. • Place and compact fill to construct dams and the spillway. • To reduce the flow rate from the trap, line the outlet with riprap and gravel over the dewatering pipe, if necessary. • The spillway weir for each temporary sediment trap should be at least 4 feet long for a 1-acre drainage area and increase by 2 feet for each additional drainage acre added, up to a maximum drainage area of 5 acres.
Monitoring	• Install monitoring posts in the trap which mark ½ the design depth for sediment accumulation. • Inspect at least once per seven calendar days, or within a reasonable period (not to exceed 48 hours) of a rainfall event which causes stormwater runoff to occur on-farm.
Maintenance	• Remove trash accumulation; remove sediment accumulations once sediment reaches half of the design depth, as indicated on monitoring posts. • Repair and re-vegetate any erosion damage. • Repair settlement, cracking, piping holes, or seepage at the embankment. • Remove after upstream areas are stabilized.

2.19 SEDIMENT BASINS

A sediment basin is a temporary pond with appropriate control structures built on a farm to capture eroded or disturbed soil that is washed off during rainstorms. The basin is designed to protect neighbouring properties from damage; and to protect the water quality of nearby streams, rivers, lakes, and wetlands. Some sediment basins are converted to permanent stormwater control practices following the completion of construction activities.

Things to take note before installing a sediment basin:

- There must be adequate space and topography for the basin to be constructed and for it to function properly.
- Sediment basins must be installed only within the property or special easement limits and where the failure of the structure will not result in loss of life, damage to homes or buildings, or interruption of use or service of public roads or utilities.
- Sediment basins can attract children, and therefore can be very dangerous. Adhere to all local ordinances regarding health and safety. If fencing of basins is required, show the type of fence and its location on the soil erosion and sedimentation control plan, and in the construction specifications.
- Chemical treatment is used in addition to the sediment basin to effectively remove sediment from runoff with smaller size fractions (fine silt and clay) down to about the medium silt size fraction.
- Standing water may cause mosquitoes or other pests to breed. Therefore, pest control measures should be applied periodically

For more comprehensive design criteria, please refer to Chapter 12 of MSMA 2nd Edition, 2012.

Removal/Reduction of TN, TP, TSS at Farm

a) Turbid
b) Coagulant and Flocculent addition

2.20 MAINTENANCE OF EROSION CONTROL STRUCTURES BMPS

This is an essential aspect because it is equally important to maintain structures as its construction. A routing maintenance should be emphasized for every structure that has been installed. Generally, maintenance of these structures means evacuation of silt, sand, and gravels that settled at the bottom or sidewalls. Sediments should be removed periodically as the reduced carrying capacity of a structure and reduce efficiency. Availability of suitable equipment and the skills of an operator is an important consideration to avoid damaging the structure. Also, farm accessibility and resources allocated for maintenance operation are important factors in order to sustain the quality and durability of erosion control structures of BMPs.

2.21 AGRICULTURAL WASTE MANAGEMENT

Waste management refers to a process of disposing of all unwanted materials on the farm. Composting is regarded as a sustainable waste management practice that converts any volume of accumulated organic waste into a usable product. When organic wastes are broken down by microorganisms in a heat-generating environment, the waste volume will be reduced. Thus many harmful organisms are subsequently destroyed. A potentially useful product will then be produced (Figure 2.30). Organic wastes may include manure from livestock operations, animal bedding, and yard wastes, such as leaves and grass clippings, and even kitchen scraps. Guidelines are sometimes provided to ensure environmentally sound management practices is achieved.

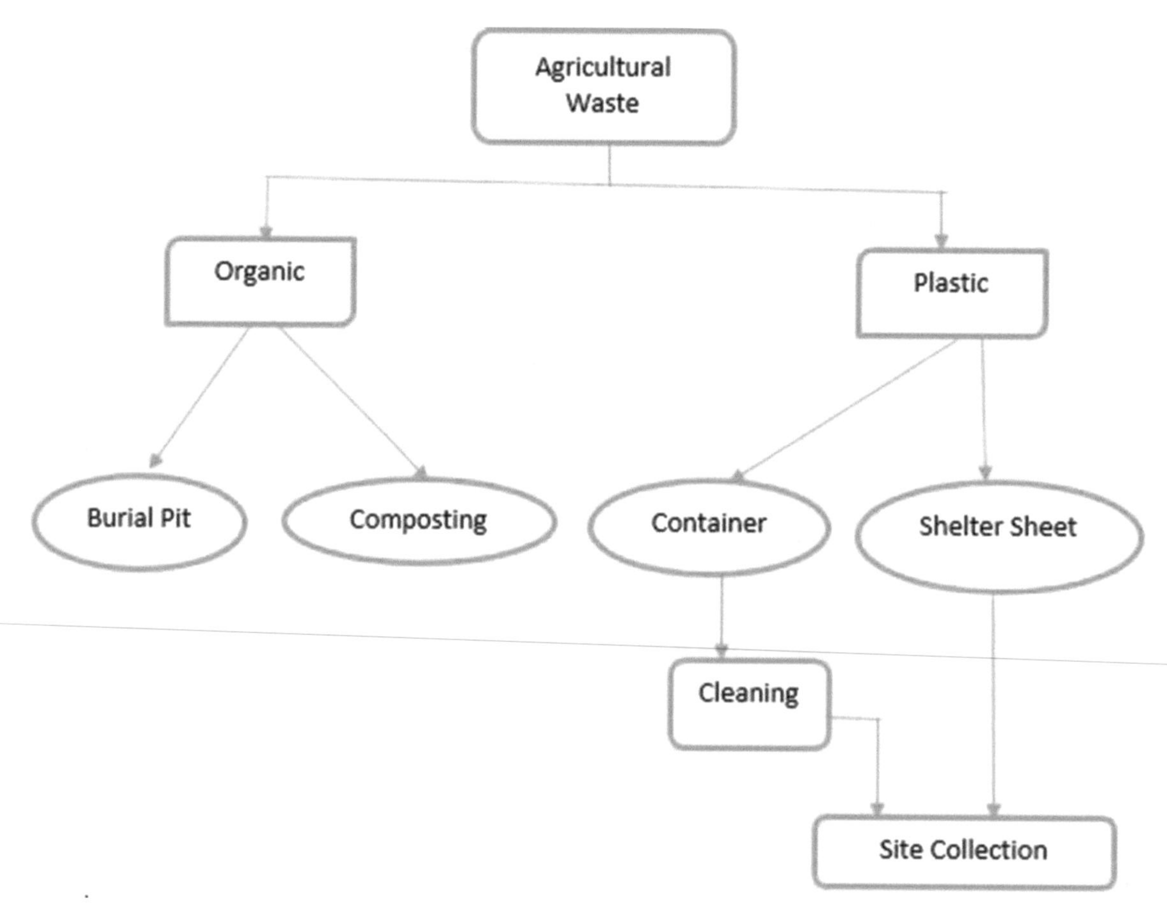

Figure 2.31: Agricultural Waste Management

Agricultural Waste management for BMPs

- Many hazardous materials are not suitable for composting. A small amount of an unsuitable product can destroy a large amount of compost. Make sure all harmful materials are removed completely
- When grass clippings are added to the compost pile for increasing N content, the lawn should be chemical-free, otherwise, plants receiving the compost may be seriously damaged.
- Plants with especially damaging diseases, such as late blight of tomato and potato, which is caused by the fungus *Phytopthera*, should not be composted to avoid the spread of the disease.
- Pesticides can only be composted if it is clear from the label that the material in question will break down into harmless components in the composting process and will not kill the composting microorganisms. Pesticide labels should list proper disposal methods.
- Plastics cannot be composted and must be recycled or disposed of in a landfill for BMPs. Tires, metals items which cannot be separated according to specific content (aluminium, steel, etc.), and plastics are difficult to dispose of and hard to manage. It is possible to find businesses which will recycle tires and sorted metals.
- Meat and other animal products may be composted in some situations. In a large scale system, even large livestock carcasses may be composted. However, caution should be taken when composting animal products in home compost piles

2.22 COMPOST FORMATION FOR BMPS

Composting is a process by which organic wastes are broken down by microorganisms, generally bacteria and fungi, into simpler forms. The microorganisms use the carbon in the waste as an energy source. The degradation of the nitrogen-containing materials results in the breakdown of the original elements into a much more uniform product which can be used as a soil amendment. The heat generated during the process kills many unwanted organisms such as weed seeds and pathogens. Advantages of composting include reduction of waste volume, elimination of heat-killed pests, and the generation of useful material. Adding compost to soil increases organic matter content which improves soil structure and allows for the slow release of nutrients for crop use in subsequent years.

In Cameron Highlands, it was observed that some farms were fully organic (MyGAP MyOrganic farms) and they do recycle waste product. Figure 2.31 shows one of the farms forming compost using biodegradable farm waste products.

Figure 2.31: Compost Formation in Cameron Highlands

Chapter III:

PREPARATION OF AGRICULTURAL EROSION AND SEDIMENT CONTROL PLAN (ESCP)

3.1 GENERAL REQUIREMENTS OF ESCP

An agricultural erosion and sediment control plan (ESCP) should include all of the applicable best management practices to remove sediments from surface runoff that flows from tilled areas directly into perennial streams.

3.1 CONTENT OF AN ESCP

A conservation plan consisting of several documents such as the conservation plan map, soil map, digital map, and the inspection and maintenance plan shall be submitted for evaluation to the relevant authorities for approval.

3.2 MAPS

Conservation Plan Map - This map shall indicate the boundaries, areas, and type of land use for each farm/fields, and the permanent conservation measures currently installed and/or planned. A title block on the conservation plan map shall include the name of the farmer, location and area of the farm, the date the map was prepared, the name of the person who prepared the map and the map scale. The map shall include a legend that explains all of the symbols used in the map.

Soil Map - The soil map shall possess information about the type of soils present within the designated field/farm area and its boundaries. A title block on the soil map shall include the name of the farmer, the location and area of the farm, the source of the soil survey data and the map scale. The map shall include a legend that explains all of the symbols used, especially the types of soil shown on the map. A description of the mapping units, which consists of the soil parent material, slope, texture, drainage characteristics, available moisture, depth, erosion hazard, and fertility characteristics shall be included.

Digital Map - The digital map shall indicate the precipitation rate, slope characteristics, and vulnerability towards a landslide. The risk assessment is carried out initially using the digital map.

3.3 REPORTS

Conservation Plan Reports – The report shall include the names, units, amount and location of each of the conservation practices already installed or to be installed at the farm/field. Implementation of the conservation plan means that the various measures shall be installed in sequential order. For example, a waterway may serve as an outlet for diversions. Thus, the waterway must be installed, stabilized and be in operating condition before the diversions are constructed. The date (year) when the existing practices were installed and a schedule (month and year) for practices not yet installed must be included in the report.

3.4 CONSTRUCTION DRAWINGS

Other information to be added to the Conservation Plan - Before the installation of structural conservation practices, detailed designs and construction drawings for the practices are to be prepared. Details of interim erosion control measures used during the installation of these structural measures shall be included in the construction drawings.

3.5 ESCP PREPARATION STAGES

The ESCP must be prepared for land clearing, planting season and fallow period phases for open, rain-shelter and terraced plot (Figure 3.1).

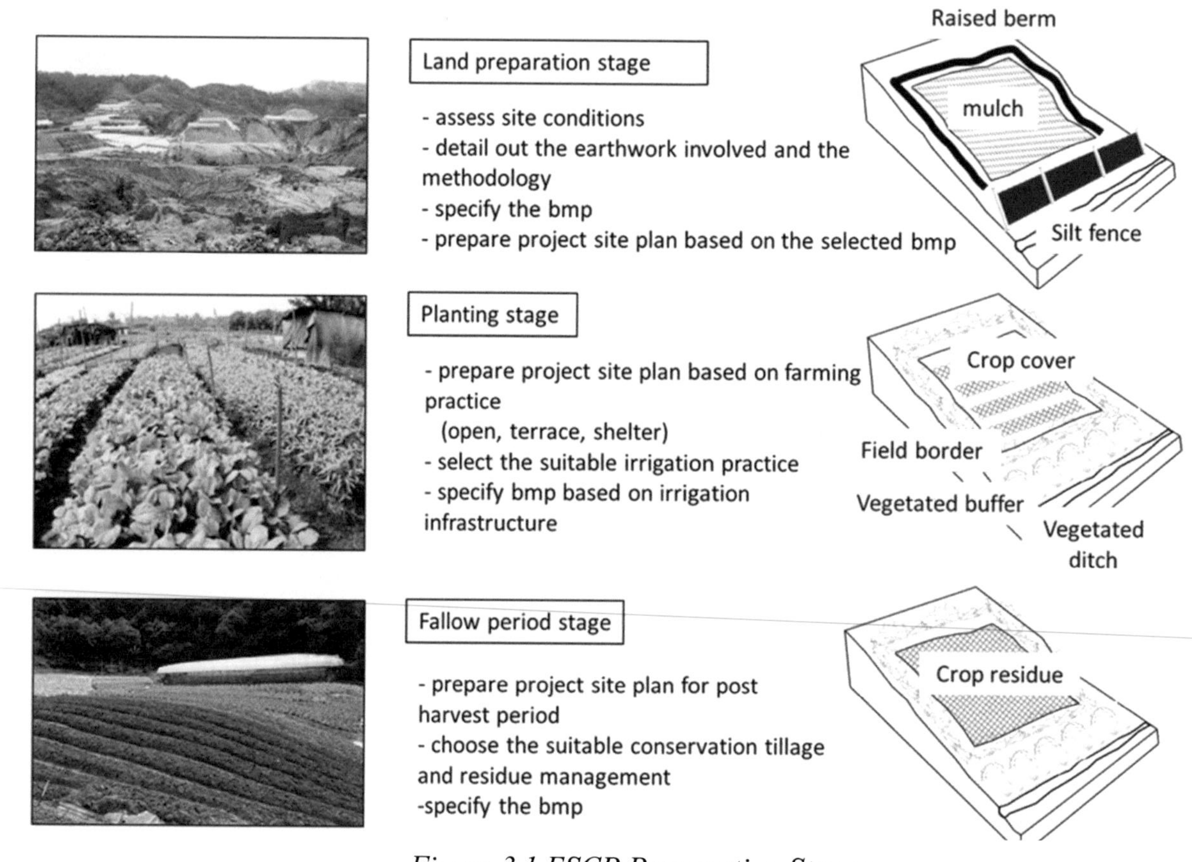

Figure 3.1 ESCP Preparation Stages

3.6 LAND CLEARING PHASE

There are various land-clearing methods, namely; manual, mechanical and chemical. The most important component of this phase is the prevention of soil erosion and sedimentation.

Temporary Conservation Cover - Establishing and maintaining perennial vegetative cover to protect soil and water resources on the land. This will help to reduce soil erosion and sedimentation, thus protecting water quality in addition to improving wildlife habitat.

Access Road Construction - An access route constructed as part of a conservation plan. Its purpose is to provide a fixed route for travel for moving livestock, produce, equipment, and supplies. It may also be used to provide access to proper operation, maintenance and management of conservation enterprises and conservation farming systems while controlling runoff to prevent erosion.

Conservation Tillage System - Any tillage and planting system in which at least 30 per cent of the soil surface is covered by plant residue after planting to reduce soil erosion by water during the critical period.

Grade Stabilization Structure - A structure to control the grade and head cutting in natural or artificial channels. A waterway or engineered channels need stabilizing structures if the surrounding slopes are very steep, causing the velocity of runoffs to exceed the safe limits provided by vegetation. Energy dissipaters may be required at the outlets of the structures. The structures shall be designed and constructed to provide permanent stabilization. Types of grade stabilization structures include straight drop spillways, box inlet drop spillways, drop boxes, concrete chute spillways and drop inlet spillways.

Sediment Basin - A basin constructed to collect and store sediment and debris. This practice applies where conditions preclude the installation of erosion control measures from keeping soil in place or where a sediment basin offers the most practical solution to containing erosion.

3.7 PLANTING SEASON PHASE

Critical Planting Area - Planting vegetation such as trees, shrubs, vines, grasses or legumes on highly erodible or critically eroding areas (does not include tree planting mainly for wood products).

Roof Runoff Management (Rain shelter) - A facility for collecting, controlling and disposing of runoff from roofs. This helps to prevent runoff water from flowing across concentrated waste areas, barnyards, roads, and alleys, and to reduce pollution and erosion, improve water quality, prevent flooding, improve drainage, and protect the environment.

Field (open) - Growing crops in a systematic arrangement of strips or bands across the general slope (not on the contour) to reduce water erosion. The crops are arranged so that a strip of grass or a close-growing crop is alternated with a clean-tilled crop or fallow. This practice will reduce sheet and rill erosion and improve surface water quality by reducing siltation.

Terrace - An earth embankment or a ridge and channel constructed across the slope at a suitable spacing and with an acceptable grade. These terraces shall be built to reduce erosion and sediment content in runoff water. They can also be used to improve farming ability, reduce the risk of flooding, and increase soil moisture.

Runoff Management System - A system for controlling excess runoff caused by construction operations at development sites, changes in land use, or other land disturbances. This is done to minimize such undesirable effects as flooding, erosion, sedimentation and to maintain or improve water quality.

Diversion - A channel constructed across the slope with a supporting ridge on the lower side. This is done to divert excess water from one area for use or safe disposal in other areas. Establishment and maintenance of permanent stabilization (usually grass) are required.

Field Border - A band of grass or legume planted at the edge of a field. This practice is used to control erosion and to protect edges of fields that are used as turn rows of travel lanes

Cover and Green Manure Crop - A crop of close-growing grasses, legumes or small grain grown primarily for seasonal protection and soil improvement. It usually is grown for one year or less to control erosion during periods when major crops do not furnish adequate cover.

Mulching - Applying plant residues or other suitable materials (not produced on the site) to the surface of the soil. This application will help conserve moisture; prevent surface compaction or crusting; reduce runoff and erosion; control weeds, and help establish plant cover.

Filter Strip - A strip or area of vegetation for removing sediment, organic matter and other pollutants from runoff. This is done via several different processes such as filtration, infiltration, absorption, adsorption, decomposition, and volatilization.

Sediment Basin - A basin constructed to collect and store sediment and debris. This practice applies where conditions preclude the installation of erosion control measures from keeping soil in place or where a sediment basin offers the most practical solution to containing erosion.

3.8 FALLOW PERIOD PHASE

Conservation Tillage System - Any tillage and planting system in which at least 30 percent of the soil surface is covered by plant residue after planting to reduce soil erosion during the critical period.

Conservation Cover - Establishing and maintaining perennial vegetative cover to protect soil and water resources on land retired from agricultural production. This will help reduce soil erosion and sedimentation, thus protecting water quality and improving wildlife habitat.

Crop Residue Use - Using plant residues to protect cultivated fields during critical erosion periods. This practice shall be planned to reduce soil loss during the critical periods, to conserve soil moisture, to maintain or improve soil tilth, to minimize fluctuations in soil temperature, and to provide food and cover for wildlife.

3.9 PERFORMANCE EVALUATION OF ESCP

3.10 SITE INSPECTION & MONITORING

Monitor the erosion and sediment control measures and maintain, repair, adjust, and supplement them as construction and weather conditions require. Provide for regular inspection and maintenance in every erosion control plan, including the following:

- Designate inspection tasks and frequencies and a responsible party to perform inspections and follow-up adjustments and repairs.
- Designate a responsible party to monitor the weather.
- Post erosion and sediment control plan drawing in the
- Project field office.
- Document changes to the erosion and sediment control plans. Changes can be handwritten and strikeouts in the original document.
- Monitor and replenish stored erosion control supplies and equipment to respond to the occurrence of unforeseen weather conditions and erosion and

- Sediment problems.
- Store records of inspection, maintenance, corrective action at the site

Inspection Reports - After every inspection, an inspection report should be completed and include:

- Summary of the areas inspected
- The name(s) and title(s) of personnel making the inspection
- The date(s) of the inspection
- Major observations
- Corrective actions are taken
- Any incidents of non-compliance

3.11 RECORD KEEPING

Proper record-keeping is very important to save information for BMPs. The information is normally kept in diaries, inspection and maintenance reports, meeting minutes, and photos are crucial for documenting a defence of "due diligence" whenever there is landslide occurrence. Records can be valuable proof that a proper ESC plan was implemented, adequately monitored and maintained. At a minimum, the owner or developer must complete an inspection report of any maintenance, damages or deficiencies of erosion and sediment control measures. An inspection should be undertaken and prepared the report once per week and following heavy rainstorms or snowmelt events. The same document can be used to record maintenance and repairs undertaken following an inspection. It is the responsibility of the owner or developer to prepare the inspection report. The inspection report must be signed by the owner's/developer's inspector.

References

ASABE Standards (2011). Standard Engineering Practices Data. Adopted and Published by American Society of Apicultural and Biological Engineers.

Briggs Irrigation Manual (1991). Boyle Road Corby Northampton shire, England http://www.briggsirrigation.co.uk/products/tied-ridger/

Erosion and Sediment Control Guidelines, Canada (2005). Stantec Consulting Ltd. under contract to The City of Edmonton, Drainage Services.

Frederick R Troeh (2003) Soil and Water Conservation for Productivity and Environmental Protection. ISBN-13: 9780130968074

DOA, 2013. Garis Panduan: Pembangunan Pertanian Di Tanah Bercerun. Cetakan Pertama (2013) Edisi Pertama. ISBN 978-983-047-196-9

Hamzah, Z., Amirudin, C. Y., Saat, A., & Wood, A. K. (2014). Quantifying Soil Erosion and Deposition Rates in Tea Plantation Area, Cameron Highlands, Malaysia Using 137 Cs. The Malaysian Journal of Analytical Sciences, 18(1), 94–106. Retrieved from http://www.ukm.my/mjas/v18_n1/Che Yasmin.pdf

ILO and UNDP, (1993) Special Public Works Programs - SPWP – Planting Trees - An Illustrated Technical Guide and Training Manual 190 p. http://dfsc.dk/Extensionstudy/069%20Planting%20Trees/B1010_13.HTM.

Kelvin K. K. Kuok1*, Darrien Y. S. Mah2, P. C. Chiu3, (2013). Evaluation of C and P Factors in Universal Soil Loss Equation on Trapping Sediment: Case Study of Santubong River. Journal of Water Resource and Protection, 2013, 5, 1149-1154. 1149 – 1154 p

Manual Seliran Mesra Alam Malaysia (MSMA). Arbun Storm Water Management Malaysia. Pusat Penyelidikan Kejuruteraan Sungai dan saliran Bandar (REDAC). Kampus Kejuruteraan, University Sains Malaysia. Seri Ampangan 14300, Nibong Tebal, Plau Pinang. (2012) 2nd Edition

Planning and Design of Drainage in Hilly Area (2012): A Conceptual Guideline by Integrated land use planning and water resources management (ILPWRM). A centre sponsored by Ministry of Urban Development, Govt. of India.

Raj, J. K. (2002). Land-use changes, soil erosion and decreased the base flow of rivers at Cameron Highlands, Peninsular Malaysia. Bulletin of the Geological Society of Malaysia, Annual Geological Conference 2002 Issue, 45, 3–10.

Seaboard (2017). Farming Programs system in Eastern Seaboard. Programs to Help Aspiring Farmers Bring Vision to Fruition. https://www.naturalblaze.com/2017-/08/farming-programs-on-the-eastern-seaboard.html.

Sediment & Erosion Control Plans (SECP). Curriculum Study Guide. An Irrigated Lands Regulatory Program Grower Self-Certification Training. Central Valley Water Quality Coalitions

Smith D.D. and Whitt, D. M., Zingg, Austin W., McCall, AG. And Bell. F.C. (1947). Investigation in erosion and reclamation of eroded Shelby and related Soils and Conservation Expt. Sta., Bethany, Mo., 1930 – 1942. Washington, D.C., US. Department of Agriculture. Technical Bulletin No. 883

Teh, S. (2011). Soil erosion modeling using RUSLE and GIS on Cameron Highlands, Malaysia for hydropower development, 76.

Williams, J. (1975). Sediment Yield Prediction with Universal Equation using Runoff Energy Factor. Agricultural Research Service Report ARS-S-40, U.S. Department of Agriculture.

APPENDIX A: EXAMPLE: SOIL LOSS ESTIMATION

1 Reference	2 Calculation	3 Output
4	5 Universal Soil Loss Equation (USLE) will be used to assess the erosion risk of the site under three conditions, i.e. existing (undisturbed), disturbed and uncontrolled (no ESC), and disturbed but controlled (with ESC). 6 7 Hence; 8 9 USLE parameters for the area erosion index of L12 (Appendix F) under the existing condition. The procedures to obtain USLE parameters for other zones and development conditions are precisely the same as that shown in this example. 10 **1) Determination of Rainfall Erosivity, *R* Factor:** *11* 12 *R* factor for the area of Kea Farm (Cameron Highlands) falls in the range of 15,000 to 17,500 MJ.mm/ha.ha.yr. For evaluation purpose, the higher limit is used, therefore, *13* 14 *R* Factor = 17,500 MJ.mm/ha.hra.yr 15 **2) Determination of Soil Erodibility, *K* Factor :** 16 17 In determining the K factor of the developed area, soil data obtained from hand auger method for the site is used. 18 19 The soil samples are tested for grain analysis, and the results are converted to 100% of sand, sit, clay and organic matter (excluding larger particles), as K = 0.0244 20 **3) Determination of Slope, LS Factor:** 21 22 when slope length = 100m; and steepness = 30%; 23 24 Therefore LS = 18.046 25 **4) Determination of C values:** 26 27 For agricultural area, C = 0.38 could be adopted.	**33**

1 Reference	2 Calculation	3 Output
	28 5) **Determination of P values:** 29 30 Under terracing condition and slope of more than 20%, 31 32 P = 0.18 can be adopted.	
34	35 Thus, **36** **37** Soil Loss, **A = R.K.LS.C.P** 38 39 A = 17,500 x 0.0244 x 18.046 x 0.38 x 0.18 **40**	**41** **42** **43** **44** **45 527 ton/ha/yr**

APPENDIX B: EXAMPLE: SLOPE S HEIGHT COMPUTATION

46 Reference	47 Calculation	48 Output
49 From table	50 *Example:* Calculate the ultimate shear strength of a soil 6 m below the surface, giving that, the soil density is 124 kg/m^3. Assume the soil is a well-graded type. 51 52 Solution. 53 For well-graded sand, Φ = 32-35° = 33.5° 54 55 Normal stress = (124 kg/m^2) (6 m) = 744 kg/m^3 56 tan $\Phi = \tau_f / \sigma$; Hence; 57 $\tau_f = \sigma * \tan \Phi$ 58 τ_f = 744 kg/m^2 * tan(33.5°) = 492 kg/m^2 59	60 τ_f = 492 kg/m^2 61
62 Example:	63 Using the following information to calcite soil shear strength. Make assumptions where necessary. 64 65 Given: Cohesion strength = 500 kg/m^2 66 Unit weight = 110 kg/m^3 67 Slope steepness = 50° 68 Internal friction angle = 15° 69	70 θ =15°
71 Example:	72 To calculate the design foot bearing load (q_{design}) in kg/m^2 of a slope with the following given parameters; 73 74 Given data: ➢ Strip footing 3 m wide ➢ Wet soil with density of 125 kg/m^3 ➢ Angle of internal friction = 30° ➢ Cohesive strength of 400 kg/m^2 ➢ Use safety factor of 3 75 Solution: 76 a_1 = 1.0, 77 a_2 = 0.5, 78 B = width = 3' 79 γ_1 = 125/2 = 62.5 kg/m^3; 80 γ_2 = 125 kg/m^3 81 c = 400 kg/m^2 82 N_c = 30, 83 N_γ = 18, 84 N_q = 20	90 Design foot beating, 91 q = 7900 kg/m^2

46 Reference	47 Calculation	48 Output
	85 86 $q_{ult} = (1.0)\left(\frac{400kg}{m^2}\right)(30) + (0.5)(3m)\left(\frac{62.5kg}{m^3}\right)(18) + \left(\frac{125kg}{m^3}\right)(4m)(20) = 23,700\, kg/m^2$ 87 88 $d_{esign} = q_{ult} / FS$ = 23,700/3 = 7900 kg/m² 89	
92 Example:	93 To calculate the maximum slope height of a soil for the information given below; 94 Given data: ➢ Cohesion strength = 500 kg/m² ➢ Unit weight = 110 kg/m³ ➢ Slope steepness = 50° ➢ Internal friction angle = 15° 95 96 The maximum slope height could be determined as follows; 97 98 Solution; 99 From the chart (b) in Figure 5.3, 100 the values of φ = 15°, i = 50° 101 102 Hence, 103 104 H_{max} = c / (γ * N_s) 105 = (500 kg/m²) (1m³/ 10) (1/ 0.095) 106 = 4.8 m	107 H_{max} = 4.8 m

APPENDIX C: EXAMPLE: DESIGN OF BUNDS TERRACING FOR OPEN FARMLAND AS A CONTROL MEASURES

108 Reference	109 Calculation	110 Output
111	112 113 <u>Determination vertical intervals of a bund</u> 114 115 A bund to be constructed in heavy rainfall zone and on a sloppy land of 25^0, the; 116 117 The vertical interval of bund, 118 119 VI = 30 x Slope/3 + 60 120 121 VI = 30 x0.25/3+60 122 = 62.5 cm 123 = 0.625 m	124 VI = 0.625 m @ high rainfall areas like that of Cameron Highlands
125	126 If the plot above were located in low rainfall zone, the vertical bund interval would be; 127 128 VI = 30 x slope/2 + 60 129 = 30 x 0.25/2 + 60 130 = 63.75 cm 131 = 0.638 m	132 VI = 0.638 m @ low rainfall areas
133 Equation 2.8	134 <u>Determination of Vertical Intervals (VI)</u>; 135 136 For an area inclined on a slope of 30° with rainfall and infiltration capacity of 3 mm and 1 mm respectively taking Manning's coefficient and maximum permissible velocity of 0.001 and 0.5 m/s. 137 138 Then, the slope length; 139 140 $L = \frac{V^{5/2} n^{3/2}}{(R-i)\sin^{3/4}\theta\cos\theta}$ 141 142 $L = \frac{0.5^{5/2} 0.001^{3/2}}{(3-1)\sin^{3/4} 30\cos 30}$	145

	143 144		
146 Equation 2.9	147 <u>Vertical interval VI</u> 148 VI = L Sin θ 149 150 = L x sin 30° 151	152	

APPENDIX D: EXAMPLE: STORMWATER MANAGEMENT OF SHELTERED FARM

153 Reference	154 Calculation	155 Output
156 Equation 2.8	157 <u>The Required drain size</u> 158 159 The drain pipe diameter for a storm with 6 mm intensity could be computed as; assuming flow velocity of 160 161 v = 2 m/s 162 D =√ 6/(2π) /30 = 0.10233 m = 3.99 inch pipe	163 Drain size, d = 4 inch
164	165 For 20m x 50 m sheltered farm, the total area covered is $1000m^2$ 166 167 Assuming, rainfall with an intensity of 3 mm occurred, 168 169 The total volume of runoff is 1000 m^2 x 3 10^{-3} m 170 = 3 m^3/hr	171

APPENDIX E: MAPS OF SOIL EROSION AND LANDSLIDES SUSCEPTIBILTY

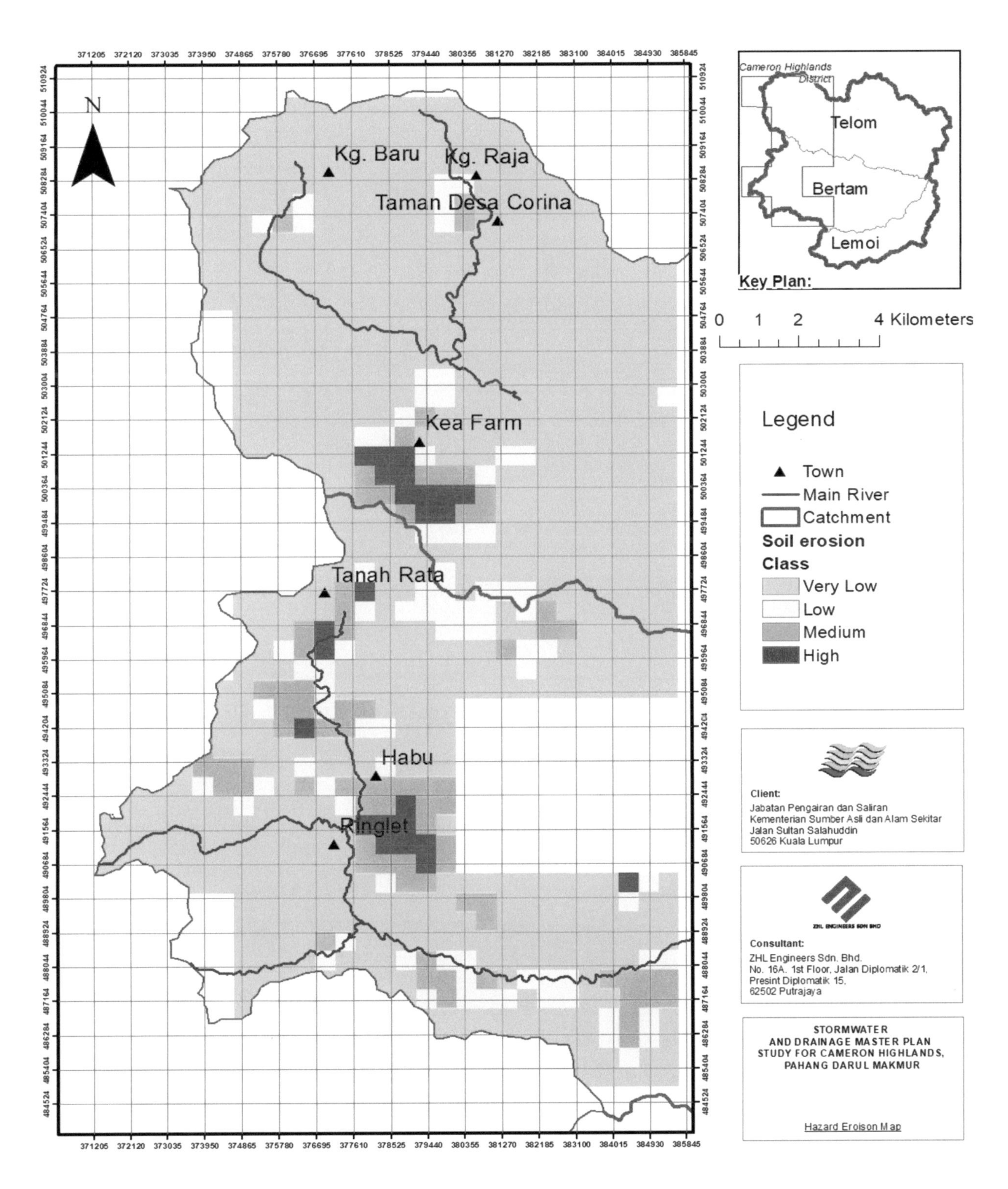

CAMERON HIGHLAND SOIL EROSION INDEX

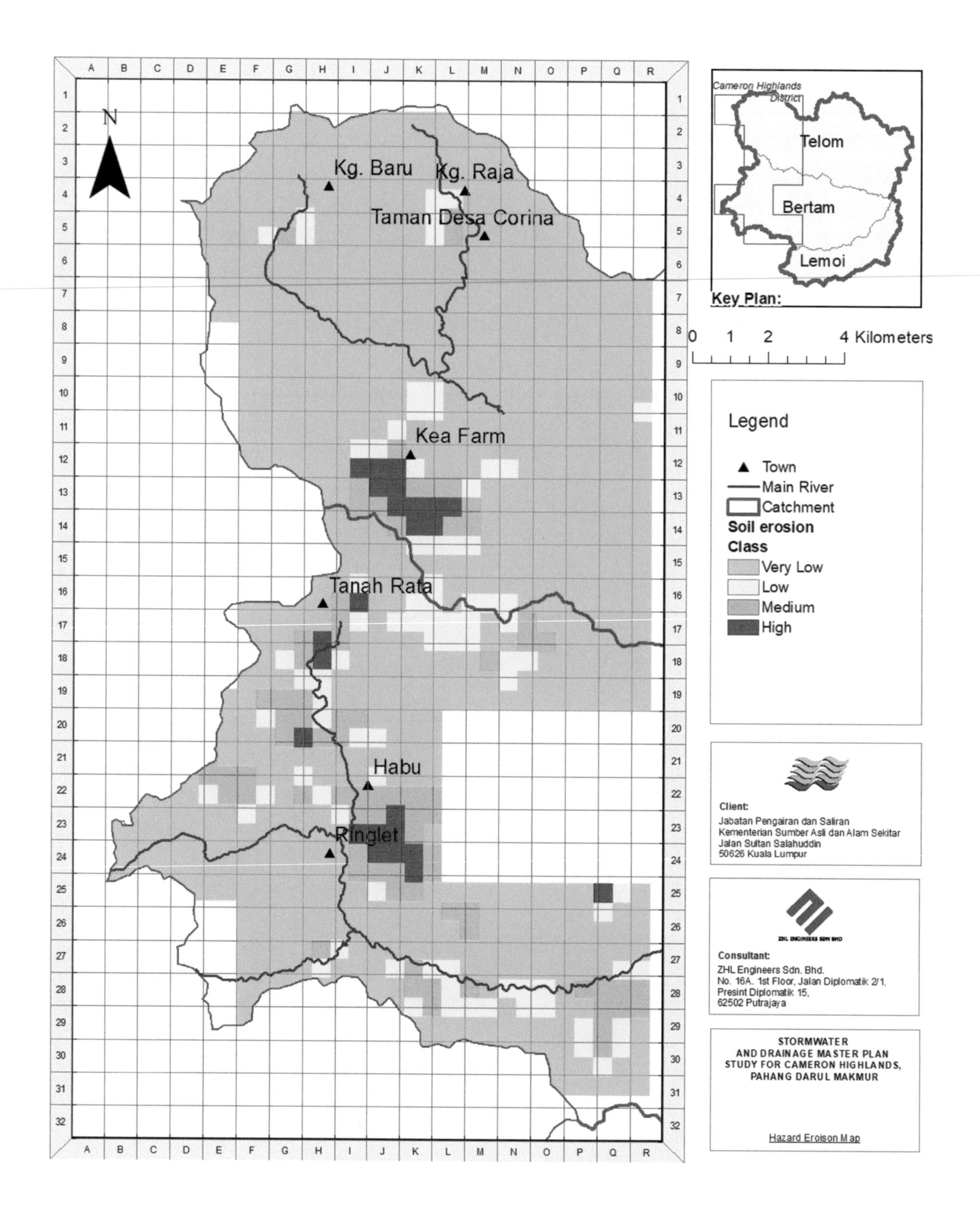

CAMERON HIGHLAND
LANDSLIDE SUSCEPTIBILITY MAP

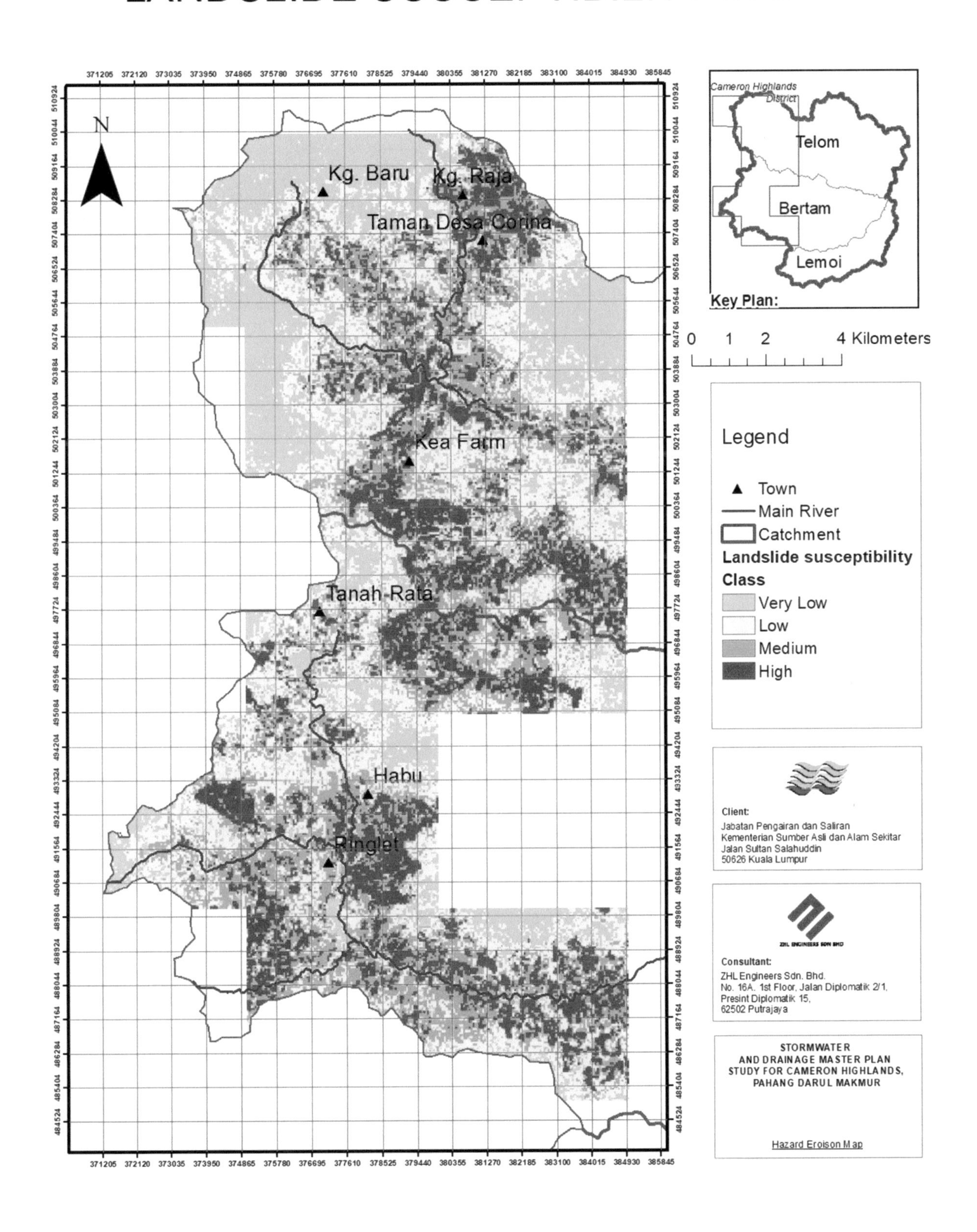

CAMERON HIGHLAND
LANDSLIDE SUSCEPTIBILITY INDEX

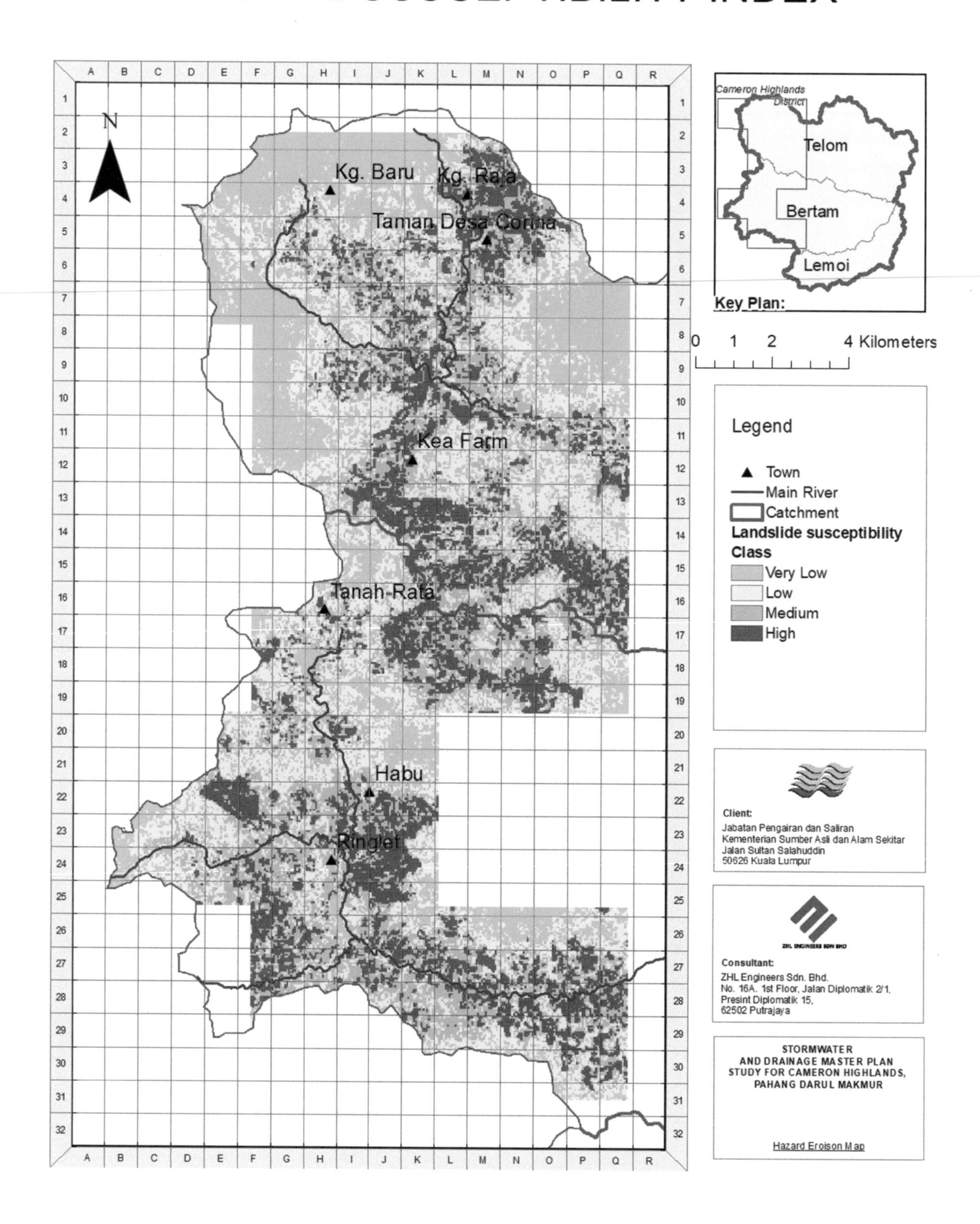

CAMERON HIGHLAND
EROSION INDUCED LANDSLIDE MAP

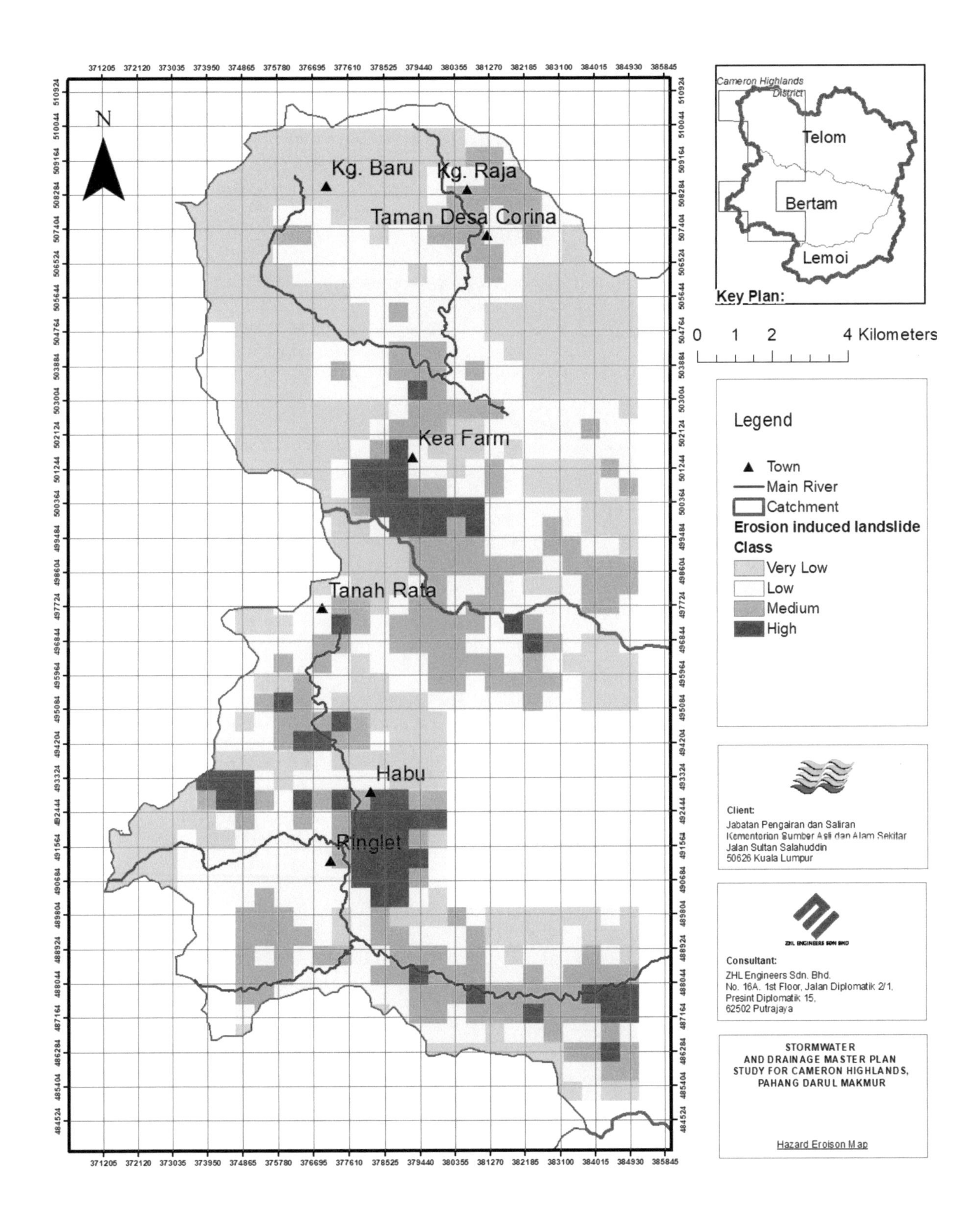

CAMERON HIGHLAND
EROSION INDUCED LANDSLIDE INDEX

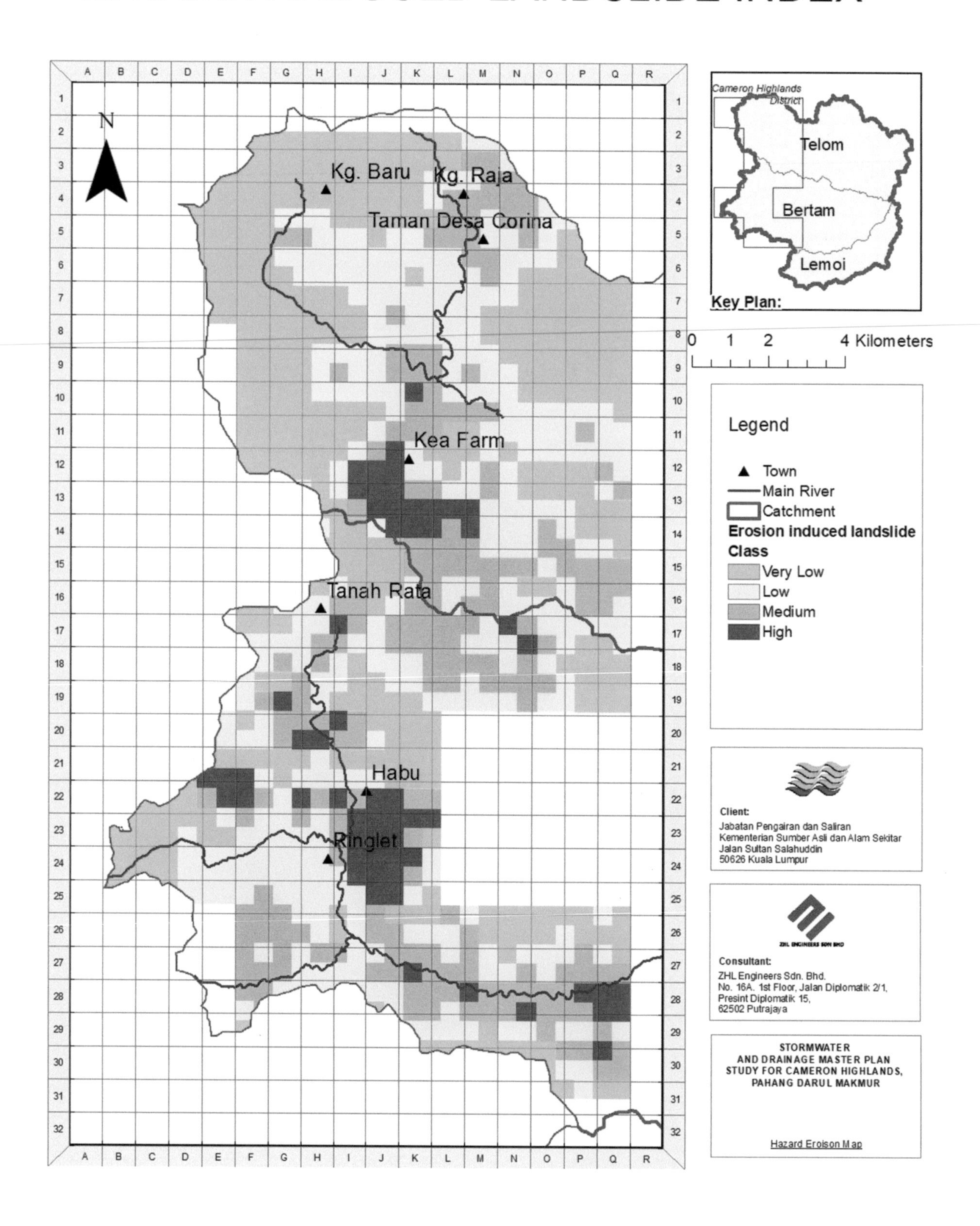

172 **EROSION AND SEDIMENTATION CONTROL AT AGRICULTURAL FARM CHECKLISTS**	173 174 175 176 **DEPARTMENT OF IRRIGATION AND DRAINAGE**
177	
178 **PART I: PERSONAL INFORMATION**	
179 Name: 180 181	182 I.C. No.:
183 Date of Birth: Age: 184 185	186 Farm Location: 187 188 189 190
191 Gender: □ Male □ Female 192	
193 Tel (H): 194 195	196 Home Address: 197 198
201 Tel (O) 202	199 200
203	
204 **PART II: AGRICULTURAL INFORMATION**	
205	
206 □ New Farm □ Existing Farm 207	
208 1. Farm Type: 209 210 1.1. Open Farm: □ Terracing □ Contouring 211 212 1.2. Shelter: 213 214 1.2.1 □ Shelter on leveled ground 215 □ Shelter on the terrace 216	221 Crop 1: Area size: 222 223 Slope....................................degree 224 □ Open farm □ Sheltered farm 225 226 Crop 2: Area size: 227 228 Slope....................................degree 229 □ Open farm □ Sheltered farm 230

217 1.2.2 □ Soil culture 218 □ Hydroponic culture 219 □ Fertigation 220	231 Crop 3: Area size: 232 233 Slope....................................degree 234 □ Open farm □ Sheltered farm 235 236 237 238 239 240

241 2. Major drainage system:

242

243 Distance to river..m River size..................................m

244

245 3. Access road facilities:

246

247 Distance to the main road..km

248

249 Road type..

250

251 Size..m

252

253 4. Farm Facilities:

254

255 4.1 □ Worker's house

256

257 4.2 □ Chemical storage (fertilizer, pesticide, herbicide etc.)

258

259 4.3 □ Composting center

260

261 4.4 □ Temporary plastic waste center

262

263 4.5 Irrigation water source:...

264

265 Distance to the source..m

266

267 4.6 Irrigation water storage: □ Pond □ Tank; Size...m^3/litre

268

269 4.7 Irrigation Type: □ Sprinkler □ Drip □ Furrow □ Flooding

270 (If sprinkler and drip system, pipe size is..............................mm)

271

272 4.8 Drainage System Layout: □ Yes □ No

273 (If Yes, the dimension of each component is..
....
....)
274

275 5. Equipment and Vehicle:
276
277 □ Tractor □ Sprayer □ Generator □ Hand tools
278
279 type..
280
281 □ Truck/Pick-up
282
283 □ Others:...
284
285
286
287
288

289 6. Erosion Susceptibility Map:
290
291 Reference Index.........................(As in Appendix E)
292 Erosion Rate Level: □ Very Low □ Low □ Medium □ High
293

294 7. Landslides Susceptibility Map:
295
296 Reference Index.........................(As in Appendix G)
297 Landslide Risk Level: □ Very Low □ Low □ Medium □ High
298

299 8. Erosion Induced Landslides Risk Map:
300
301 Reference Index.........................(As in Appendix I)
302 Landslide Risk Level: □ Very Low □ Low □ Medium □ High
303

304 **PART III: EROSION AND SEDIMENTATION CONTROL PLAN**

305 A: OPENING NEW FARM

306

307 1. Tree removal: □ Tree cutting □ Burning □ Export
308

309 2. Land clearing method: □ Manual □ Mechanical □ Chemical
310
311 If Mechanical; states the heavy machinery types: ....
....
312

313 3. Land reclamation: □ Yes □ No
314
315 3.1 Reclamation area size.......................................m^2
316 (Cutting area..............................m^2; Landfill area................................m^2)
317
318 3.2 Height of cutting slope.......................................m
319
320 If YES;
321 Layout plan for cut and fill submission: □ Yes □ No
322

323 4. Period of opening a new farm (months):...;
324 Monsoon Season: □ Yes □ No
325

326 5. ESCP during the opening of new farm:
327
328 5.1 Map of general topography (slope gradients, lengths, orientation): □ Yes □ No
329
330 5.2 Soil types (grain size, erodibility):...
331
332 5.3 Layout/map of drainage patterns – provide topography map with contour intervals
333 sufficient to show drainage patterns, drainage divides, and flow directions:
334 □ Yes □ No
335
336
337
338 5.4 Landslide protection measure at the cut slope:
339 □ Gabion □ Retention wall □ Crop cover/Turfing
340 □ Terracing (Terrace angle.....................degree; Height........................m)
341 □ Others...
342
343 5.5 Erosion and sediment control measure at the site:
344 □ Silt fence □ Side drain □ Sediment trap □ Sediment basin □ Others................
345
346 5.6 Construction of terrace/contour:
347
348 5.6.1 Terrace/contour type:...
349
350 5.6.2 Erosion and sediment control measure for terrace/contour farm:
351 □ Side drain □ Crop cover/Turfing □ Sediment trap
352 □ Sediment basin □ Others..
353
354 5.7 Erosion and sediment control measure along the access road:
355 □ Side drain □ Crop cover/Turfing □ Sediment trap □ Sediment basin
356 □ Others...

357
358 5.8 Location and description of permanent stormwater management facilities including storm
359 drain inlets, pipes, outlets, waterways, swales, ponds, etc. :
360 □ Yes □ No
361
362 5.9 Sediment monitoring system at the drainage outlet:
363 □ Yes □ No
364
365 5.10 Submission of farmstead plan layout:
366 □ Yes □ No
367

368 B: LAND PREPARATION

369 1. Open farm:
370 1.1 Map of general topography (slope gradients, lengths, orientation): □ Yes □ No
371
372 1.2 Duration:..........................; Monsoon Season: □ Yes □ No
373
374 1.3 Layout/map of drainage patterns – provide topography map with contour intervals
375 sufficient to show drainage patterns, drainage divides, and flow directions:
376 □ Yes □ No
377 1.4 Soil types (grain size, erodibility):..
378
379 1.5 Maintenance plan for terrace/contour – provide the information of the inspection of the
380 planting bed, vertical interval, ridges, mulching, side drain, etc.: □ Yes □ No
381
382 1.6 Maintenance of planting bed and vertical interval of the terrace:
383 □ Mulching □ Crop management □ Ridges □ Side drain
384 □ Others..
385 1.7 Soil tillage operation:
386 □ Zero tillage □ Rotary tillage □ Plough tillage □ others..
387 1.8 Location and description of permanent storm water management facilities including storm
drain inlets, pipes, outlets, waterways, swales, ponds, etc.: □ Yes □ No
388 1.9 Rain water harvesting plan layout and description: □ Yes □ No
389 1.10 Sediment monitoring system at the drainage outlet: □ Yes □ No
390 1.11 Waste management plan:
391
392 1.11.1 Waste handling: □ Separation process □ Burial pit □ Composting
□ Deliver to the collection site
393
394 1.11.2 Distance to Waste Collection Center....................km
395
396 1.11.3 Road Accessibility: □ Yes □ No
397 If any; the road condition is...
398

399 2. Sheltered farm:
400
401 2.1 Number of rain shelter:……………………………………………………………
402
2.2 Submission of plan layout and dimension of rain shelter: □ Yes □ No
403
2.3 Maintenance of terrace bed and its vertical interval (if applicable):
404 □ Mulching □ Crop management □ Perimeter drain
405 □ others……………………………………………………………………………
406
2.4 Rain water harvesting plan layout and description: □ Yes □ No
407
408 2.5 Location and description of permanent stormwater management facilities including
409 gutter, downspout, pipes, outlets, perimeter drain, waterways, swales, ponds, etc. :
410 □ Yes □ No
411
2.6 Sediment monitoring system at the drainage outlet: □ Yes □ No
412
413 2.7 Waste management plan:
414
415 2.7.1 Waste handling:
416 □ Separation process □ Burial pit □ Composting □ Deliver to the collection site
417
418 2.7.2 Distance to Waste Collection Center…………………km
419
420 2.7.3 Road Accessibility: □ Yes □ No
421 If any; the road condition is……………………………………………………
422
423
424

425 C: PLANTING SEASON

426 1. Crop type:……………………………………………………………………………
427

2. Farming practices:
428 2.1 Open farm – erosion and sediment control measures:
429 □ Mulching □ High density cropping □ Strip cropping
430 □ Terracing with strip cropping
431 □ Others……………………………………………………………………………
432
2.2 Sheltered farm - : □ Maintenance of rain water harvesting
433 □ Maintenance of stormwater management facilities
434

3. Irrigation systems layout and design description –(to prevent runoff water and soil erosion): 435 □ Yes □ No 436
4. Waste management plan: 437 438 4.1 Waste handling: □ Separation process □ Burial pit □ Composting 439 □ Deliver to the collection site 440 441 4.2 Distance to Waste Collection Center..............................km 442 443 4.3 Road Accessibility: □ Yes □ No 444 If any; the road condition is.. 445
446
447 D: FALLOW PERIOD
448
1.1 Erosion and sediment control measure after harvest: □ Yes □ No 449 450 If yes; by □ Mulching □ Crop cover □ Others.. 451
1.2 Maintenance of planting structures during fallow period: 452 1.2.1 Maintenance of planting terrace –replacement of plastic mulch, planting bed, etc.: 453 □ Yes □ No 454 1.2.2 Maintenance of rain shelter- replacement of shelter plastic: □ Yes □ No 455 1.2.3 Maintenance of stormwater facilities: □ Yes □ No 456
1.3 Waste management plan: 457 458 1.3.1 Waste handling: □ Separation process □ Burial pit □ Composting 459 □ Deliver to the collection site 460 461 462 1.3.2 Distance to Waste Collection Center...............................km 463 464 1.3.3 Road Accessibility: □ Yes □ No 465 If any; the road condition is... 466

467 FOR OFFICE USE ONLY:	
468 Date of Submission: 469	470 Review Date:
471 472 Submission: Acceptable/Rejected 473 474 Date:	475 476 By:... 477 478 Name:

Printed in the United States
By Bookmasters